SpringerBriefs in Electrical and Computer Engineering

More information about this series at http://www.springer.com/series/10059

Leonardo Gabrielli · Stefano Squartini

Wireless Networked Music Performance

Leonardo Gabrielli
A3Lab
Università Politecnica delle Marche
Ancona
Italy

Stefano Squartini
Ancona University
Ancona
Italy

ISSN 2191-8112 ISSN 2191-8120 (electronic)
SpringerBriefs in Electrical and Computer Engineering
ISBN 978-981-10-0334-9 ISBN 978-981-10-0335-6 (eBook)
DOI 10.1007/978-981-10-0335-6

Library of Congress Control Number: 2015958535

Printed on acid-free paper

This Springer imprint is published by SpringerNature
The registered company is Springer Science+Business Media Singapore Pte Ltd.

Preface

Networked music performance is often intended as a practice of remote performance, with musicians displaced far away at distant locations. Networking, however, does not only have the potential of getting remote performers closer, but also allows local performers to get farther, explore larger spaces, move to outdoor venues, yet feel as close as they need to perform well. Furthermore in computer music and electroacoustic practice, live electronics can be digitally interconnected in unprecedented ways, thanks to networking. The laptop performance paradigm already exploits networking for control data exchange. However, with the steady switch from analog links to digital connections, possible due to improvements and adoption of digital audio networking technologies, creative possibilities for performers are open to explore and the exploitation of wireless links is the necessary step to really make this possibilities of convenient adoption.

A comprehensive state of the art in networked music performance (NMP) and a historical survey of computer music networking is reported to later introduce current technical trends in NMP and technical issues are yet to address. Wireless communication protocols are listed and compared to the requirements of NMP. Two chapters are, then, dedicated to works related to wireless NMP, especially those from the authors, who have devoted their last 3 years on wireless music networking research. The current outcomes of the WeMUST project are reported with practical results and application reports.

The last chapter closes the book, with future issue to investigate. Lastly, an appendix providing a short addendum on wired Audio over IP standards is given.

Ancona
June 2015

Leonardo Gabrielli
Stefano Squartini

Acknowledgments

The authors would like to acknowledge Adelmo De Santis, Michele Bussolotto, and Fons Adriaensen for providing useful support in audio and wireless networking development. The authors would also like to acknowledge the role and vision of Paolo Bragaglia in providing a direction for the development of WeMUST and supporting its live performance use during festival Acusmatiq. Many thanks to the first performers who employed the WeMUST system and provided feedback, Giampaolo Antongirolami, Laura Muncaciu, and Enrico Francioni.

Contents

Abbreviations

ACL	Asynchronous Connection-Less
A/D	Analog-to-Digital conversion (also, ADC)
AES	Audio Engineering Society
ALSA	Advanced Linux Sound Architecture
API	Application Programming Interface
BSS	Basic Service Set
CSMA/CA	Carrier Sense Multiple Access with Collision Avoidance
CSMA/CD	Carrier Sense Multiple Access with Collision Detection
DAW	Digital Audio Workstation
DCF	Distributed Coordination Function
DFA	Delay Feedback Approach
DHCP	Dynamic Host Configuration Protocol
DLL	Delay-Locked Loop
DNS	Domain Name System
DSP	Digital Signal Processor or Digital Signal Processing depending on the context
D/A	Digital-to-Analog conversion (also, DAC)
DSL	Digital Subscriber Line
EPT	Ensemble Performance Threshold
FFT	Fast Fourier Transform
FTA	Fake Time Approach
GP	General Purpose
GPS	Global Positioning System
GUI	Graphic User Interface
IC	Integrated Circuit
ICN	Information-Centric Network
IP	Internet Protocol or Intellectual Property depending on the context
ISM	Industrial Scientific Medical
JACK	Jack Connection Kit
LAA	Latency-Accepting Approach
LAN	Local Area Network

LBA	Laid Back Approach
LTE	Long-Term Evolution
LUT	Look-Up Table
MA	Musical Application
MAC	Multiply and ACcumulate
MIDI	Musical Instruments Digital Interface
MIMO	Multiple Input Multiple Output
MPC	Mixing Personal Computer
MSA	Master–Slave Approach
NFC	Near Field Communication
OS	Operating System
OSC	Open Sound Control
OSS	Open Sound System
PCF	Point Coordination Function
PDF	Probability Density Function
PDV	Packet Delay Variation
PLL	Phase-Locked Loop
PLR	Period Loss Rate
QoS	Quality of Service
RIA	Realistic Interaction Approach
RISC	Reduced Instructions Set Computing
RTT	Round-Trip Time
SABy	Simple Autonomous Buddying
SCO	Synchronous Connection-Oriented
SFG	Signal Flow Graph
SoC	System on a Chip
SME	Small-to-Medium Enterprise
SNR	Signal-to-Noise Ratio
SP	Signal Processing
STP	Standard Temperature and Pressure
TCP	Transmission Control Protocol
UDP	User Datagram Protocol
UPnP	Universal Plug and Play
WAN	Wide Area Network
WASN	Wireless Acoustic Sensor Network
WLAN	Wireles Local Area Network
WPS	Wi-Fi Protected Setup

Chapter 1
Introduction

The scope of this book is to provide a quick and informed introduction to the issues and developments in Networked Music Performance (NMP) to, then, introduce the use of wireless technologies for transmission of audio (and possibly video) signals in short- to medium-range contexts, which is a very recent development of NMP. The topic stands at the intersection among several research areas including audio and video digital signal processing, wired and wireless communications, psychoacoustic research, and musical aesthetics. While computing and networking shall be treated at length, given the technical nature of this book, some basic aesthetic and psychoacoustic concepts for NMP are also reported together with a brief history, in order for the reader to become acquainted with the main concepts that drive composers and performers to explore NMP and remote telepresence for musical performance. The feasibility studies and the developments carried out by the authors, in the context of the WeMUST project, are reported at length. A few practical examples of music performance and art installations employing wireless networking are described. Given the recent introduction of wireless networking in music performance, a whole chapter is devoted to issue yet to solve and future trends that can be foreseen for research in the field in the years to follow.

The outline of the book follows

- Chapter 2 gives a definition and a taxonomy regarding networked music performance. A brief historical overview is provided together with current experiences of networked laptop orchestras and remote networked music performance;
- Chapter 3 introduces challenges and technical issues imposed to the music performance by networking;
- Chapter 5 introduces recent developments in wireless music performance and studio practice, their outcomes, and the solutions found to some of the technical issues reported in the previous chapters;

L. Gabrielli and S. Squartini, *Wireless Networked Music Performance*,
SpringerBriefs in Electrical and Computer Engineering,
DOI 10.1007/978-981-10-0335-6_1

- Chapter 6 reports details on the use of wireless NMP in real-world contexts;
- Chapter 7 concludes the book.
- An Appendix is provided which reports an overview of audio over IP standards.

Chapter 2
Networked Music Performance

Abstract Networked music performances have been considered since a few decades in contemporary music composition. More recently, the enabling technologies for NMP have been also considered as a way to deliver classes to facilitate rehearsal or conduct proper music improvisation. Unfortunately after a few years of experiments through the Internet the interest has shifted toward the use of NMP for composition only and advancement in avant-garde music practice. NMP is also a means of reflection for the artists and the composer over new media technologies. For these reasons planning a NMP requires an interdisciplinary approach, evaluating both aesthetic and technical issues. In this book, intended for the technical reader, only the latter shall be addressed. However, to ease the exchange and expedite collaboration between technologists and artists an introduction on historical and current practices in NMP is provided. The lexicon and the dissertation on technical issues are intentionally left as simple as possible to facilitate readers with no specific technical background on digital signal processing, audio processing, and networking.

Keywords Networked music performance · Network computer music · Laptop orchestra

2.1 Definition and Taxonomy

One of the first definitions of networked music performance dates back to 2001 [1]:

> A Networked Music Performance occurs when a group of musicians, located at different physical locations, interact over a network to perform as if they would, if located in the same room.

Such definition carries an implicit bias, as the network is seen as a surrogate for the natural propagation of sound and light in the space between musicians sharing the same ambience. Other researchers, e.g., those of the SoundWire group at Stanford CCRMA, showed rather the opposite interest, i.e. to explore the network as a new medium for interaction and making its inherent shortcomings, such as the propagation delay, as a feature, or at least something musicians have to "*live with*" [2].

L. Gabrielli and S. Squartini, *Wireless Networked Music Performance*,
SpringerBriefs in Electrical and Computer Engineering,
DOI 10.1007/978-981-10-0335-6_2

A more neutral and ample definition of NMP is thus,

> the practice of conducting real-time music interaction over a computer network.

This definition, thus, does not suggest limitations on the way the interaction is conducted, the distance, or the instruments to be employed. Many category of NMP fall under such a broad definition. We may thus build a taxonomy of networked music performances based on the instruments employed: *computer-only*, *human performers-only*, or *mixed*. We shall see later in Sect. 2.2 that the first usage of networking was conducted between computers employing algorithms for sound generation, with limited human intervention. We shall also see that in the recent years fiber-optic networks for research institutions allowed simultaneous low-latency transmission of audio and video, making it possible for acoustic musicians displaced at remote location to perform and interact in a feedback fashion. One reason why computer-only NMP is first seen in NMP history is that it can be carried out by transmitting control data only, or some form of compressed representation of audio data, thus it has been feasible since the inception of computer networking, given the limited bandwidth available at the time. NMPs can be, thus, classified whether the network carries *control data only*, *audio data only*, or both. Computer networking is often employed in the *laptop orchestra* paradigm. Technical and musical discussion of laptop orchestras is a topic on its own, which is already treated by textbooks and reviews. Not all laptop orchestras employ networking. Indeed, they seldom employ networking and most of the times it is based on TCP/IP data exchange over a wired connection.

Another means to discriminate between different NMP practices is the distance and position of the performers and their instruments. Most NMP literature at the moment deals with remote performers, i.e., performers located at distances much larger than 1 km. In this book, the use of wireless technologies enables, instead, to connect musicians located in the same room, in a large indoor space or in outdoor spaces. We may, thus, distinguish between *indoor local*, *outdoor local* or *remote* NMP. In the former two cases local area networking (LAN) technologies are employed, while in the second wide area networking (WAN) is necessary. The level of interaction also defines different kinds of performance: the musicians can be *tightly synchronized*, as they are when they improvise in a shared space, they can be *loosely synchronized*, i.e. aware of each other's actions, but not able to respond as they were separated by a negligible latency or *disconnected*. In the latter case, they may be disconnected *aurally* or *visually*, or both. If there is not aural or visual connection a click track or similar means must be employed to synchronize them (Table 2.1).

To conclude, there are a number of different approaches for musical interaction between human performers, depending on latency constraints, proposed by Carôt in [3]:

- Realistic Interaction Approach (RIA),
- Master Slave Approach (MSA),
- Laid Back Approach (LBA),
- Delay Feedback Approach (DFA),

Table 2.1 Taxonomy of a networked music performance

Human role	Instrument performer, laptop performer, supervisor only (autonomous computer network)
Network topology	Start, point-to-point, mesh
Transmitted signals	Audio, video, control (e.g. OSC), text chat
Distance of performers	Remote (tens to hundreds of km), outdoor local (up to a few km), indoor
Networking area	LAN, MAN, WAN
Networking technologies	Wired LAN, wireless LAN, fibre-optic WAN, copper WAN, satellite link
Latency and synchronization	Tight synch, loose synch, click synch (no aural/visual cue), disconnected (only the audience is aware of the performers actions)
Audio and video	Aurally and/or visually synchronized, aurally and/or visually aware but not synchronized, no aural and/or visual connection

- Latency Accepting Approach (LAA),
- Fake Time Approach (FTA).

The **RIA** is the most demanding, as it tries to simulate the conditions of a real interplay with the musicians in the same space. The general latency threshold for this approach is set by Carôt at 25 ms for the one-way delay (or latency), following an early technical report from Nathan Schuett at Stanford in 2002 [4], although more accurate studies exist in literature which suggest slightly different (but comparable) values. In the **MSA**, a master instrument provides the beat and the slave synchronizes on the delayed version that is coming from the master. The audio is in sync only at the slave side, while the master does not try to keep up with the slave's tempo, but he can barely get a picture of what is going on at the slave's side, trying not to get influenced in his tempo by the incoming delayed signal. Clearly, the interaction gets reduced in this approach, but the acceptable latency increases. The **Laid Back Approach** is based on the laid back playing manner, which is a common and accepted solo style in jazz music. Playing laid back means to play slightly behind the groove, which musicians often try to achieve consciously in order to make their solo appear more interesting and free. Similarly to the MSA, at the master side the beat is built and at the slave side, a solo instrument can play. At the master side, the round trip delay, if higher than 25 ms but below 50 ms, creates an artificial laid back style of playing. LBA of course does not work for unison music parts in which both parties have to play exactly on the same beat at the same time. A commercial software Musigy exist that implements LBA. A **DFA**, tries to fill the latency gap between the two ends by introducing an artificial delay in the listening room at the master end (if one of the ends have a master role), or at both ends (if interplay is not hierarchical). In the first case, e.g., delaying the master's signal in the room allows to make it closer to the slave's delayed signal. Similarly for the non-hierarchical case. The approach however, introduces a latency between the user action (e.g., key press) and aural response (in case of an electronic instrument) or simply adds a delayed version to an acoustic instrument, which can deteriorate playing conditions, however, is suitable with turntables or sound sources with little human interaction. A commercial software, eJamming, employs this approach. The **LAA** simply neglects synchronization and is used for contemporary avantgarde music, music with very low timing constraints or computer music which employs the network as part of the performance. The SoundWire group promoted this approach with several performances of contemporary music.

Finally, an approach that accepts latency but allows for tempo (but not beat) synchronization is the **FTA**. In this case the latency is artificially adapted to be one measure or multiples. This way, any performer plays on the previous measure executed by the other performer. This approach requires a tempo to be known a priori and fixed. A further hypothesis is needed, that the music does not change drastically from measure to measure, which is the case for many improvisational genres, such as blues, funk, etc. The Ninjam open source software employs this approach. For a different, more aesthetics-related classification of NMP, refer to [5].

2.2 A Brief Timeline

Transmitting music over a distance by technological means was accomplished at the outset of telephone and radio technologies, for technological display and entertainment. One of the first wired music transmission was possible due to Thaddeus Cahill, inventor of the Telharmonium (patent no. 580035, 1896), an "apparatus for generating and distributing music electrically," which since 1906 was employed in performances and could drive 15,000–20,000 telephone receivers, according to its main capital investor, Oscar T. Crosby [6]. Its demise was soon to come,[1] as radio broadcast was going to spread in the subsequent years. And annoyed mistresses were not ready for electronic music yet.

The first radio broadcast dates back to 1910, with an experimental transmission of Mascagni's *Cavalleria Rusticana* and Leoncavallo's *Pagliacci* featuring the Italian tenore Enrico Caruso from the Metropolitan Opera House, in NYC, USA. The amplitude-modulated broadcast, provided by the American inventor Lee De Forest's Radio Telephone Company had some issues, especially with the scarce microphone signal loudness, but the path was set. In Fig. 2.1 a New York Times advertisement for the radio (called *wireless*, despite De Forest's preference for the former term) is

[1] Cable radio was anyway employed all along the twentieth century, with a varying degree of success from country to country and different developments. Its main advantages over radio transmission are a more capillary reach, a higher quality compared to AM transmission and reduced costs compared to digital transmission. Music broadcasting services by means of cable radio are still in use in several countries including Italy.

Fig. 2.1 An advertisement of *wireless music* well before the WeMUST project

shown. Although other musical wireless transmission were experimented before,[2] the one from 1910 was the first to address a public.

Despite the long history of music broadcasting, music performance over a distance has a much more recent history. Radio modulated waves had an influence over John Cage, who is credited by some authors to be the first composer for a NMP piece [7], with his "Imaginary Landscape no. 4 for Twelve Radios" in 1951 (see also [8]). The piece does not fall into what has been previously defined as NMP, given the absence of a network. However, disputable whether this can be considered an NMP piece, it is surely one of the first human attempt to explore musical interaction at a distance. The same author in 1992–1993, composed *Helikopter-Streichquartett* a string quartet piece to be played on four helicopters. The piece is the third scene of the opera *Mittwoch aus Licht*. The musicians are separated aurally and visually, and are synchronized by a click track and do not hear each others, while the audience can see and hear all of them through audio and video signals transmitted from each helicopter. Networking was first employed by Max Neuhaus in 1966 in his pieces for public telephone networks. In *Public Supply I*

[2]Most notably, the one from Dr. Nussbaumer at the University of Graz, in 1904, who yodeled an Austrian Folk song from a room to another by means of a receiver and transmitter.

he combined a radio station with the telephone network and created a two-way public aural space twenty miles in diameter encompassing New York City, where any inhabitant could join a live dialogue with sound by making a phone call. Later, in 1977 with Radio Net, he formed a nationwide network with 190 radio stations.[3]

2.2.1 Early Computer Network Experiments

Computer networking experiments in music were introduced by the League of Automatic Music Composers (LAMC), an experimental collective formed by San Francisco Bay area musicians John Bischoff, James Horton, Tim Perkis, and others. The production of the LAMC was improvisational and the group was open. The group recorded music in the years 1978–83. The LAMC experimented with early microcomputers, and also very primitive computer networking. Specifically, they employed multiple MOS KIM-1 microcomputers programmed by themselves in the 6502 CPU machine language, inputting the assembly by means of a numeric keypad. Programs were stored in audio cassettes. Without getting into the aesthetic and compositional aspects of the LAMC experience, it is of interest here to report on the technical details. The microcomputers were interconnected through parallel ports or interrupt signals, directly handled by the microcode written by the composers. No standard networking was employed. The interaction was based on musical representation data, later fed to analog synthesizers or direct D/A for sound synthesis. The flier from an early concert from band members Bischoff, Behrman, Horton and Gold, depicted in Fig. 2.2 shows the data path and algorithms employed during the then-upcoming event. This picture is still quite representative of laptop orchestra performances taking place nowadays. Please note that audio signals were not transmitted digitally.

In [9], Bischoff, Gold and Horton, report on another performance taking place at the same venue in July 1978, where three KIM-1 are interconnected in a different fashion and output sound through direct D/A or a 8253 programmable interval timer chip. The interconnection is made through serial or 4-bit parallel ports.

The LAMC proposed the term Network Computer Music for their performances. Such performances provided the basis for the typical laptop orchestra paradigm, with different units exchanging musical or audio data which are subsequently processed by other units for synthesis or manipulation. Notably, the setup would algorithmically generate music and sound in a deterministic but unpredictable way, thanks to the feedback nature of the system. The composers themselves claimed to be influenced by some of the intellectual currents of the time, suggesting that complex phenomena could emerge from the interconnection and interaction of simple components. By the way, it is worth citing some of these scientists and writers since their writings were necessary to much computer music theory and practice, still influencing laptop orchestras and NMP composers: Ilya Prigogine (complex system theory and self-organizing systems), Warren S. McCullough (Neural Networks), Gregory

[3]From Max Neuhaus official webpage.

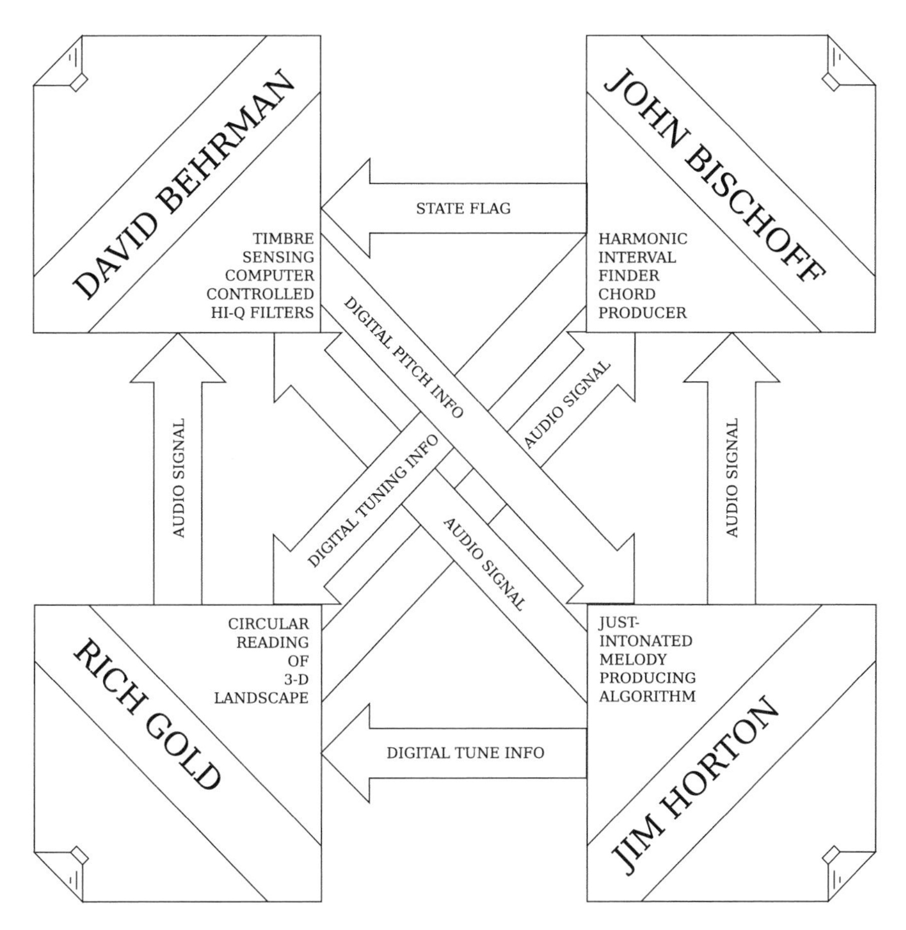

Fig. 2.2 Reproduction of a flier publicizing a concert at the Blind Lemon in Berkeley, USA in Nov. 1978. The flier features a diagram of the data paths between computers and indicates the musical algorithm running at each microcomputer

Bateson (cultural ecology), John Holland (genetic algorithms). More introspection on the LAMC can be found on a document written by Perkis and Bischoff available on the Internet "The League of Automatic Music Composers 19781983."[4] After the LAMC stopped its activities in 1983, due to Jim Horton's health problems, Perkis and Bischoff followed on the same path and worked on a digital system interconnection interface for musical practice called *The Hub*, which eventually led to the creation of a stable group of performers with the same name. The KIM-based Hub had four UARTS to allow four players to network using 300 bps serial connections. This central unit allowed for an easier connection (during the LAMC days the computers were connected by directly wiring and soldering the machines) and standardized

[4]http://www.newworldrecords.org/uploads/fileIXvp3.pdf.

format for data exchange. The musicians would, in fact, employ a shared memory (nicknamed "The Blob") where to store data. This was referred to as "blackboard system" and allowed for asynchronous data exchange. The hub would keep information about each player's activity accessible to other players' computers. As reported in [5], after the 1985 Network Muse festival held in San Francisco, featuring Bischoff, Perkis and more artists, *The Hub* collective would develop and add distance to their networked performances. In a series of concerts in 1987 six musicians would play in two different venues in New York City, split into two groups, connected through a telephone line via a modem. The technical effort was considerable, but successful, although one of *The Hub* members, Gresham-Lancaster in [10] comments

> Although the group performed at separate locations a few times, it created its strongest and most interesting work with all the participants in the same room, interacting directly with each other and with the emergent algorithmic behavior of each new piece

With the advent of MIDI and software sequencers, in 1990, Gresham-Lancaster and Perkis designed a MIDI-based Hub, where each musician's machine was assigned a MIDI port and MIDI messages were employed to exchange data. This way each musician could address privately each other machine, adding flexibility. At the end of the 1990s *the Hub* members explored the use of the Internet as a means to develop their network computer music experiments. Again, the comments by Gresham-Lancaster are not entirely positive,

> In the only test case so far, two of us performed from each of three sites [...]. This formidable test actually ended up being more of a technical exercise than a full-blown concert. [...] In this case, the technology was so complex that we were unable to reach a satisfactory point of expressivity.

Later in 1997, *The Hub* musicians were asked for a new remote performance, which was called "Points of Presence," a live performance produced by the Institute for Studies in the Arts (ISA) at Arizona State University (ASU), linking members of the Hub at Mills College, California Institute for the Arts, and ASU via the Internet. Communication technology was not mature for musical usage at the time and the experience reported by member Chris Brown with networking at a distance was not adequate enough, as he explains in an article[5]:

> ... the performance was technically and artistically a failure. It was more difficult than imagined to debug all of the software problems on each of the different machines with different operating systems and CPU speeds in different cities. In part, because we weren't in the same place, we weren't able to collaborate in a multifocal way (only via internet chats, and on the telephone); and in a network piece, if all parts are not working, then the whole network concept fails to lift off the ground. We succeeded only in performing 10 min or so of music with the full network, and the local audience in Arizona had to be supplied with extensive explanations of what we were trying to do, instead of what actually happened.

Notwithstanding technical difficulties the group reformed after 2–3 years and went on to create new repertoire that took full advantage of the improvements to networking infrastructure in performances throughout the U.S. and Europe.

[5] http://crossfade.walkerart.org/brownbischoff/IndigenoustotheNetPrint.html.

Several other early NMPs took place in those years. In 1992 Jean Claude Risset and Terry Riley performed from Nice while David Rosenboom and Morton Subotnick performed from Los Angeles. Augmented pianos (Disklavier) were used with sensors and actuators allowing to capture locally and reproduce remotely the exact playing of the musician. The concert, named Transatlantic Concert [11] was based on satellite links, and thus, given the high cost, was not replicated, nor the project was further sustained. Satellite links indeed may seem worth investigating for NMP, if the economic expenses are ignored. Satellite links are able to provide, for extremely distant location a minimal number of switching nodes, and quite direct paths. Unfortunately, the satellite employed for this kind of communications are in the geostationary orbit, i.e., approximately 35700 km from the Earth surface. Approximating signal propagation by the light speed in vacuum, under the hypothesis the points to connect are in visibility with one satellite only, the delay introduced only by propagation uplink and downlink is of the order of 250 ms. This is unacceptable for RIA NMP. In copper or fibre-optic links, the path with wired Internet connections is much shorter, although orography and routing may increase the length of the path. Even though the signal propagates at slower velocity in these materials, it may be convenient to employ copper of fibre-optic backbones even for NMP over locations across the Earth. As an example, the distance between the two farthest points in Europe (Portugal–Finland) or the USA (Washington-Florida) is approximately 4500 km, i.e., 21 ms (under the hypothesis of a direct path and the signal traveling at $0.7c$). This is at least one order of magnitude lower than employing a satellite link (even though the fibre-optic or copper link is never direct, thus may be longer). For reasons of latency and economic expenditure satellite links have not been experimented in other network performances to the best of our knowledge.

2.2.2 *Current Networked Computer Music Performance Trends*

While at the beginning of the 2000s The Hub collective slowed down its activity, its founders continued to experiment. It is the case, for instance, of Chris Brown, who, among other works, employed networking to create a shared virtual instrument from two ReacTables,[6] one placed in Barcelona, Spain and the other in Linz, Austria. This experiment, *Teleson*, was premiered in 2005 and crosses the borders between computer music and NMP. Similar experiences started being common practice. Nowadays laptop orchestras and NMP are widespread. The LAMC and The Hub experiences were fundamental for the development of current computer music performance trends. Following their activities these two important lines of music research, namely the laptop orchestra paradigm and the networked music performance, were created. They are both of interest in this book as laptop orchestras may employ networking as well. Tens of music schools, universities and conservatories

[6]http://reactable.com/.

run a laptop orchestra. It is not by coincidence that many tutorial books on Computer Music, host a section on Networking or Laptop Ensembles.

The Viennese Farmers Manual in 1995 formed a multi-laptop ensemble, while MIMEO (Music In Motion Electronic Orchestra) mixed acoustic instruments and computers from 1997. An extremely interesting ensemble for the purpose of this thesis is PowerBooks_UnPlugged, a collective of varying number of musicians only employing laptops, started in 2003. Not thriving into the details of their very interesting aesthetic, a point to highlight is the use of laptop and wireless networking to share data, algorithms, and more. The ensemble performance is based on the music programming language Supercollider[7] and a software library developed with the contribution of the ensemble named *Republic*. According to ensemble member Alberto de Campo, Apple Ad Hoc wireless networking was employed, without any wireless Access Point. He also points out that

> [the network topology is] very democratic, everyone can send sounds to everyones server, everyones evaluated code gets sent to everyones History (for reuse, adaptation, etc.), everyone can send chat messages or shout messages to everyone else.[8]

The library allows each musician to *join the Republic* by manually entering a user name and a unique ID number. A GUI showing a chat and snippets of code employed by other musicians is presented. Synth definitions[9] can be shared, allocated, and played on other musicians' laptops. The ability of the musicians to *interfere* with each other coding flow to interact in the real environment and move freely makes this laptop orchestra very interesting.

In the last ten years, however, ensemble laptop performance has become extremely common in the literature and a plethora of papers dealing with the aesthetics, the composition techniques, or the software employed are published yearly on the proceedings of computer music conferences. Stable laptop orchestra are based in Princeton (PLOrk), Stanford (SLOrk), University of Colorado Boulder (BLOrk), Dublin, Huddersfield, Virginia Tech (L2Ork), Carnegie Mellon Laptop Orchestra, and many more. Among these, the PLOrk, reported the use of wireless networking in several setups [12, 13]. A 802.11g LAN was established to synchronize the devices, all running Max/MSP and ChucK applications to synthesize and process sound. The synchronization was, however, not very tight, as it was measured to be approximately 30–40 ms. The authors of the paper, however, stated that this was acceptable as the dislocation of the performers, similarly to an acoustic orchestra took a space of 40 ft (12 m), which naturally incurs in such latencies when the performers play acoustic instruments and the air propagates sound.[10] It is not clear whether the wireless network had a role similar to the conductor of an acoustic orchestra. In that case, indeed, the conductor is required to fill the lag due to the air propagation delays, by visually conducting. To address this issue some researchers are experimenting the use of visual cues in laptop ensembles [14, 15].

---

[7] http://supercollider.sourceforge.net/.

[8] From a private correspondence with the authors.

[9] A SynthDef or Synth definition is a Supercollider object that represent a signal processing units.

[10] The speed of sound at sea level is 340.29 m/s, i.e., approximately 35 ms for a space of 12 m.

In another more recent paper, the SLOrk reported the use of an IEEE 802.11n wireless access point for an indoor and outdoor performance [16]. The performance seems to have taken place in a garden with high trees, and the performers had to place the access point at an elevated position in order to guarantee good signal coverage for all the performers. No other information is provided regarding technical issues or the outcome, besides the remark that powering all the devices required some engineering. It is generally difficult to gather information regarding the technical issues that the orchestras encounter for their performances, since often they are barely mentioned, given the aesthetic focus of the papers. In the literature, technical aspects are often disregarded as computer musicians and composers prefer to focus on the artistic aspects and rely on widely adopted technologies. A greater understanding is gathered only by direct discussion with the involved subjects.

The lack of technical focus has probably led to withdrawal of challenges and opportunities offered by the new technologies, such as audio streaming between performers, the use of wireless networking, off-stage performance (with the notable exception of the aforementioned PowerBooks_UnPlugged ensemble and few others), the shift from laptop to embedded/embodied instruments and more. For more historical information on laptop orchestras refer to, e.g., [17].

2.2.3 Current Acoustic Networked Music Performance Experiences

Albeit the human interaction potential is much higher than that available in a networked laptop performance [14, 15], human–performer NMP has had relatively fewer occurrences reported in the literature. This is probably due to the greater technical and management effort in delivering a performance between remote locations. The Internet has been explored for NMP since its boom in the late 1990s. Early experiments, however, were unsatisfactory given the best-effort delivery nature of the Internet and later academic works mainly dealt with reserved fibre-optic links, as already mentioned. No standard has been proposed for the purpose of NMP and some software products were developed with dim success. At the outset of our research work some of these were tested, proving the impossibility of conducting performance following the RIA approach previously described, due to the latency introduced by the large network delay and buffering for jitter, even at distances of 10 km line-of-sight in a urban area, due to relatively large number of switching nodes. As a general purpose network, however, the Internet has been explored for compositional purposes (see e.g. [18]).

A *purist* approach to human–performer NMP is the use of high-bandwidth, low-latency fiber links allocated on purpose, which, although available at few institutions, provide much insight for research. Needless to say, the effort and challenges of this kind of NMP are numerous. First envisioned in the late 1990s [19], audio–video low-latency transmission has been experimented on both LAN and WAN since more than

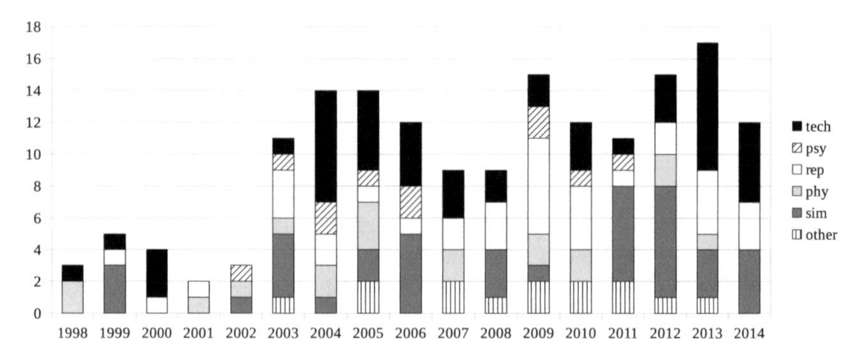

Fig. 2.3 Statistical findings from an NMP academic literature survey, years 1998–2014

a decade. Technically speaking, the first consistent attempt to address NMP over the Internet was conducted by the SoundWire group at CCRMA starting from year 2000 [20], which introduced bidirectional uncompressed audio streaming at low latency, employing the US Internet2 network. At McGill, the Ultra Video-Conferencing Research group started reporting about a similar approach, also including video, from the same year [21]. A first wave of experiments were spun in 2000 and following years. In those years the use of remote servers was often suggested (see, e.g., [22], or the Ninjam software project[11]), following similar use cases, e.g., that of remote gaming. Audio compression was also often suggested, yielding very high results in terms of latency.

In 2007, Carôt reported a decreased interest in the research communities after the first attempts in the beginning of the 2000s [7]:

> Most projects undertaken in the field of real-time high-quality networked music performance took place around and after 2000 and went very quiet afterwards. In the last year or so, a revival of interest for the subject has emerged not only on the technological side but also on the cultural side, where researchers are seeking to understand the cultural implications of providing such facilities to musicians and producers as well as seeking ways to increase the level of interactions between musicians collaborating over network connections.

To test this statement and gather further information on the literature, the authors conducted a statistical research on the corpus of academic literature from 1998 to 2014. The research, does not aim to be complete, but is thorough enough to possibly outline the trends of the NMP field since its early years up to the last year. The results are summarized in Fig. 2.3. First, academic contributions in the NMP fields have been very large from year 2003 to 2006. If we want to give account to Carôt's statement we can speculate that in years 1998–2002, the NMP was new and much hype was projected around this research field, with new projects being started. Those may have been under a phase of development around these years, and were not able to yield academic outcomes until years 2003–2006. At the time when his paper was published a decrease in NMP publications can be observed (2007 and 2008). Again, this may

[11] http://www.cockos.com/ninjam/.

not reflect the perceived interest in the academic communities, nor a decrease in the number of actual performances conducted. In the years 2008–2014 a certain degree of fluctuation can be observed.

For a deeper introspection, published papers have been separated in several distinct categories. Ph.D. degree theses, diploma theses. and master's theses have been included. The categories chosen by the authors follow:

- **tech**: technical improvements in networking, hardware, DSP, and software; new solutions and architectures for networking and management of audio and control data.
- **sim**: papers dealing with collaborative use of networks for music composition, cooperative performance also involving the audience; networking in laptop orchestras; multimodal and immersive reality for remote music performance, installations, and choreography.
- **psy**: psychoacoustic findings relevant to the field or directly targeted at NMP.
- **rep**: reports on performance experiences, reviews and surveys on performances, projects and installations.
- **phy**: philosophical works and conceptual frameworks, aesthetics dialectic of the NMP paradigm.
- **other**: development of NMP technologies for educational purposes; human–computer interfaces and visual cues; papers without specific findings or low degree of novelty.

Please note that works dealing with wireless mobile phone interface design, participatory mobile phone usage in performances, and not implying the networked performance paradigm have been discarded.

All the categories can be seen to be subject to a certain degree of random fluctuation. Specifically, by clearing out from the picture the works categorized as *sim* (Fig. 2.4), which are loosely related to NMP in that they use networking for collaborative composition and audience participation, a slightly different picture emerges, showing fewer academic outcomes in the years 2006, 2011, and 2012. The works of

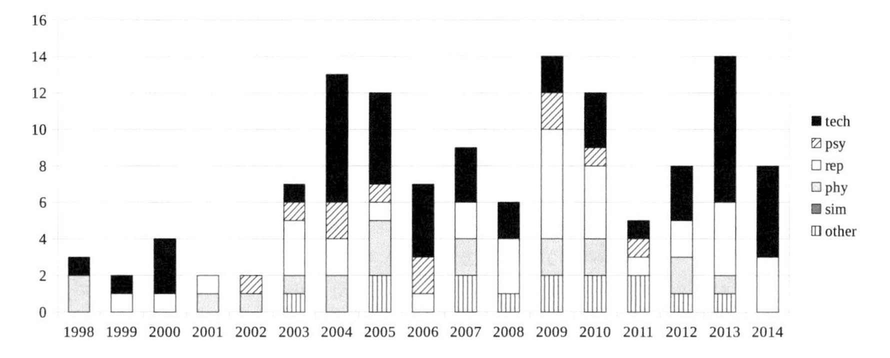

Fig. 2.4 Statistical findings from an NMP academic literature survey, years 1998–2014, showing only those works tightly related to NMP

technical nature have been numerous in the years 2004–2007, and increased again after 2011, with a peak in 2013.

In the last 5 years, a number of projects involving a large effort in technical and framework development can be seen. The DIAMOUSES framework [23], developed in Crete, provides a scalable architecture for both synchronous and asynchronous collaboration, with a web portal to share information and organize performances and the exploitation of DVB-T (digital TV broadcast) for distributed music performance on-demand replay. The MusiNet project was started at the end of 2012 and is still ongoing at the time of writing. It is funded by the European Union and involves several Greek institutions. It involves a complex network architecture involving clients and a server (*MusiNet Center*). The server can relay session information, audio/video data and further information, and textual data [24]. The choice of this architecture is motivated by a more scalable delivery, with clients responsible of sending their data to the server only and relieved of multiple transmission when more clients need to receive the same streaming. This, however, introduce additional latency. Spatial audio is taken into account by recording spatial information and estimating the direction of arrival of the sound source. Audio data is encoded with low delay and possibly acoustic echo cancelation is performed if needed. Video is also encoded, but in [24] a mechanism to reduce the latency of video streams down to 25 ms was still being investigated. This project has already provided an amount of technical outcomes related to network architectures and the use of Information-Centric Networks (ICN) [25] and Content-Centric Networks (CCN) [26] for NMP, software and cloud services for remote music collaboration [27], packet delay reduction by direct access to the network hardware [28] and more.

Differently, the LOLA project [29] aims at minimizing latencies by employing large bandwidth stable links and avoiding every step that is not essential, such as servers and compression techniques. LOLA was conceived by GARR (the Italian research institutions network) and G. Tartini Conservatory of Trieste (Italy) in 2005 and steadily developed since 2008 up to the time of writing. In LOLA, a great effort has been spent to deliver high-quality and low-latency video signals, which are interleaved and synchronized with the audio. The audio and video acquiring devices supported are limited to a small number of very high performance ones and have software drivers written to satisfy the strict time constraints. A typical setup involves a capable PC, a dedicated audio card, a USB 3.0 camera with high resolution and a gigabit network card. The monitors employed to watch the performers at the other end, must have very low latency, thus many current LCD cannot be employed given their response time of 5 ms or higher. The additional latency added to acquisition and local packet switching is the signal propagation over the fiber link, typically of the order of 1 ms per 100 km. A network administrator must guarantee at the participating institutions where the performance takes place that bandwidth is reserved for the performance. Typical bandwidth may range between 170 Mbps (640 $\times$ 480, 60 fps B/W video) to 1700 Mbps (Full HD, 30 fps color video). Video compression may be employed, reducing the bandwidth down to 20 Mbps best case, but introduces a large computational load at both ends and latency.

In the past various research projects employed video together with audio for NMP, but with very low technical demands (e.g., video link based on commercial conferencing software, of low quality and high latency, not synchronized to the audio signals). As an example, the 2008 performance of Terry Riley's *In C* between Stanford and Peking [30], made use of a 220 ms latency uncompressed audio link, while the video was streamed employing a software for multimedia movie coding and decoding (VLC) with MPEG4 compression, yielding a delay of about 1 s at 720×480 pixel resolution. The authors claimed that

> our previous experience in network performance shows that musicians usually dont look at the video when they perform; it serves primarily the purpose of providing an experience for the audience - while also adding additional reassurance and comfort to the musicians during setup, discussion and other communication needs.

This is probably in contrast with the view of the researchers behind LOLA, and a study to assess the importance of tight video requirements for the performers would be needed. Another technically challenging project in terms of video transmission is that from the McGill Ultra-Videoconferencing research group,[12] which since year 2000 started investigating the subject and in the subsequent years released a software with good results in terms of latency called UltraVideo for Linux PCs and published several works in the field [31, 32].

A similar project started in the same years is UltraGrid.[13] Ultragrid was started by CESNET, the Czech academic network operator, and SITOLA, a networking technologies laboratory established by academic partners and CESNET. The software is under active development by a team of engineers, which reportedly enables end-to-end audio and video communication at minimum latencies of 83 ms. This software is also employed for remote surgery and other uses.

Besides the few aforementioned projects focusing on technical improvement, although technical progress in communication technologies is highly relevant to NMP, most of the development efforts from the music computing research and artists communities have gone into software architectures and frameworks for NMP [23, 33, 34] or analysis and speculation of the NMP paradigm and aspects of Human–Computer Interaction [35–37]. One application close to NMP, but of different nature is music composition, editing, and collaboration by means of software and network technologies. Some of these works are, e.g., included in the *sim* category in the literature survey reported above. Only those projects (e.g., DIAMOUSES, MusiNet) and papers, that include aspects of remote music collaboration but are not limited to it, have been considered in Fig. 2.4. Collaborative composition and asynchronous NMP approaches are not considered in this book as they have different finalities and technical requirements.

A fundamental field of investigation for the feasibility of NMP is psychoacoustics. Psychoacoustic studies related to the effect of latency on synchronization and tempo keeping have been addressed by several authors [38–41] and shall be reported later

[12] http://ultravideo.mcgill.ca/.

[13] http://www.ultragrid.cz.

in Sect. 3.2. This represent a good base for the development of NMP, although many other perceptual aspects need to be evaluated in order to achieve an informed view on best practices for NMP.

To conclude, compression, coding, and error concealment DSP algorithms are covered sparsely in the literature and shall be highlighted in the respective sections.

References

1. Lazzaro J, Wawrzynek J (2001) A case for network musical performance. In: Proceedings of the 11th international workshop on network and operating systems support for digital audio and video. ACM, pp 157–166
2. Chafe C (2011) Living with net lag. In: Audio engineering society conference: 43rd international conference: audio for wirelessly networked personal devices. Audio Engineering Society
3. Carôt A, Werner C (2007) Network music performance-problems, approaches and perspectives. In: Procedings of the music in the global village-conference, Budapest, Hungary
4. Schuett N (2002) The effects of latency on ensemble performance
5. Barbosa Á (2003) Displaced soundscapes: a survey of network systems for music and sonic art creation. Leonardo Music J 13:53–59
6. Weidenaar R (1995) Magic music from the Telharmonium. The Scarecrow Press, Inc
7. Carôt A, Rebelo P, Renaud A (2007) Networked music performance: state of the art. In: Audio engineering society conference: 30th international conference: intelligent audio environments, Mar 2007
8. Pritchett J (1996) The music of John Cage, vol 5. Cambridge University Press
9. Bischoff J, Gold R, Horton J (1978) Music for an interactive network of microcomputers. Comput Music J 24–29
10. Gresham-Lancaster S (1998) The aesthetics and history of the Hub: the effects of changing technology on network computer music. Leonardo Music J 39–44
11. Polymeneas-Liontiris T, Edwards AL (2014) The disklavier in network music performances
12. Trueman D, Cook P, Smallwood S, Wang G (2006) Plork: the princeton laptop orchestra, year 1. In: Proceedings of the international computer music conference, pp 443–450
13. Smallwood S, Trueman D, Cook PR, Wang G (2008) Composing for laptop orchestra. Comput Music J 32(1):9–25
14. Rebelo P, Renaud AB (2006) The frequencyliator: distributing structures for networked laptop improvisation. In: Proceedings of the 2006 conference on new interfaces for musical expression. IRCAMCentre Pompidou, pp 53–56
15. Renaud AB (2011) Cueing and composing for long distance network music collaborations. In: Audio engineering society conference: 44th international conference: audio networking. Audio Engineering Society
16. Wang G, Bryan N, Oh J, Hamilton R (2009) Stanford laptop orchestra (slork), Ann Arbor. MPublishing, University of Michigan Library, MI
17. Collins N (2010) Introduction to computer music. Wiley
18. Mills R (2010) Dislocated sound: a survey of improvisation in networked audio platforms. In: New interfaces for musical expression, (NIME 2010), Sidney, Australia. University of Technology, Sydney
19. Bargar R, Church S, Fukuda A, Grunke J, Keislar D, Moses B, NovakB, Pennycook B, Settel Z, Strawn J, Wiser P, Woszczyk W (1998) Networking audio and music using Internet2 and next-generation internet capabilities
20. Chafe C, Wilson S, Leistikow R, Chisholm D, Scavone G (2000) A simplified approach to high quality music and sound over IP. In: COST-G6 conference on digital audio effects, pp 159–164

21. Xu A, Woszczyk W, Settel Z, Pennycook B, Rowe R, Galanter P, Bary J (2000) Real-time streaming of multichannel audio data over Internet. J Audio Eng Soc 48(7/8):627–641
22. Gu X, Dick M, Kurtisi Z, Noyer U, Wolf L (2005) Network-centric music performance: practice and experiments. IEEE Commun Mag 43(6):86–93
23. Alexandraki C, Akoumianakis D (2010) Exploring new perspectives in network music performance: the DIAMOUSES framework. Comput Music J 34(2):66–83
24. Akoumianakis D, Alexandraki C, Alexiou V, Anagnostopoulou C, Eleftheriadis A, Lalioti V, Mouchtaris A, Pavlidi D, Polyzos G, Tsakalides P et al (2014) The musinet project: towards unraveling the full potential of networked music performance systems. In: The 5th international conference on information, intelligence, systems and applications, IISA 2014. IEEE, pp 1–6
25. Stais C, Thomas Y, Xylomenos G, Tsilopoulos C (2013) Networked music performance over information-centric networks. In: 2013 IEEE international conference on communications workshops (ICC). IEEE, pp 647–651
26. Tsilopoulos C, Xylomenos G, Thomas Y (2014) Reducing forwarding state in content-centric networks with semi-stateless forwarding. In: INFOCOM, 2014 proceedings IEEE. IEEE, pp 2067–2075
27. Vlachakis G, Karadimitriou N, Akoumianakis D (2014) Using a dedicated toolkit and the cloud to coordinate shared music representations. In: The 5th international conference on information, intelligence, systems and applications, IISA 2014. IEEE, pp 20–26
28. Baltas G, Xylomenos G (2014) Ultra low delay switching for networked music performance. In: The 5th international conference on information, intelligence, systems and applications, IISA 2014, July 2014, pp 70–74
29. Drioli C, Allocchio C (2012) LOLA: a low-latency high quality A/V streaming system for networked performance and interaction. In: Colloqui informatica musicale, Trieste
30. Cáceres J-P, Hamilton R, Iyer D, Chafe C, Wang G (2008) To the edge with China: explorations in network performance. In: 4th international conference on digital arts, ARTECH 2008, Porto, Portugal, pp 61–66
31. Olmos A, Brulé M, Bouillot N, Benovoy M, Blum J, Sun H, Lund NW, Cooperstock JR (2009) Exploring the role of latency and orchestra placement on the networked performance of a distributed opera. In: 12th annual international workshop on presence, Los Angeles
32. Bouillot N, Cooperstock JR (2009) Challenges and performance of high-fidelity audio streaming for interactive performances. In: Proceedings of the 9th international conference on new interfaces for musical expression, Pittsburgh
33. Schiavoni FL, Queiroz M, Iazzetta F (2011) Medusa-a distributed sound environment. In: Linux audio conference, pp 149–156
34. Allison J (2011) Distributed performances systems using HTML5 and Rails. In: 26th Annual conference of the society for electro-acoustic music in the United States (SEAMUS)
35. Holland S, Wilkie K, Mulholland P, Seago A (2013) Music interaction: understanding music and human-computer interaction. Springer
36. Bowen N (2013) Mobile phones, group improvisation, and music: trends in digital socialized music-making, PhD thesis, The City University of New York
37. Weinberg G (2005) Interconnected musical networks: toward a theoretical framework. Comput Music J 29(2):23–39
38. Chafe C, Gurevich M (2004) Network time delay and ensemble accuracy: effects of latency, asymmetry. In: 117th audio engineering society convention. Audio Engineering Society
39. Chafe C, Caceres J-P, Gurevich M (2010) Effect of temporal separation on synchronization in rhythmic performance. Perception 39(7):982
40. Driessen PF, Darcie TE, Pillay B (2011) The effects of network delay on tempo in musical performance. Comput Music J 35(1):76–89
41. Kobayashi Y, Miyake Y (2003) Analysis of network ensemble between humans with time lag. In: SICE 2003 annual conference, Aug 2003, vol 1, pp 1069–1074

Chapter 3
Technical and Perceptual Issues in Networked Music Performance

Abstract Several NMP categories have been described in the previous chapter. The most technical demanding approaches are of special interest in this book. Low-latency audio data transfer through networking is not easily accomplished. Many research efforts generally rely on dedicated bandwidth networks for audio (and often video) transmission, such as the Internet2 in the US and the GEANT network in Europe. These are fiber networks connecting selected institutions (such as Universities) with high bandwidth (generally >1 Gbit) and low latency, by manually routing the signals, reducing the number of switches in the path, and assigning a high QoS to the audio/video signals. Reference implementations of high-quality and low-latency NMPs are based on such networks and experimental software has been developed by several research groups. Nonetheless, NMP practice still requires effort in its development, planning, and deployment phases and all technical issues must be addressed with care. This chapter of the book is devoted by introducing all the most relevant technical issues, giving a basic understanding to a broad audience ranging from musicians with a basic scientific training to audio engineers.

Keywords Latency · Dropout · Blocking delay · Network delay · Delay jitter · Remote clock synchronization

3.1 Dropout

Audio dropouts, i.e., loss of audio packets, must be avoided by any means, since a regular audio flow is the mandatory requirement for audio transmission. Source of audio dropouts are manifold:

- loss of packets along the route
- late arrival of packets at the receiving end
- loss of transmitter and receiver audio clock synchronization
- failure in the scheduling of the audio capture or playback process.

Loss of packets is often solved in networking applications by triggering a new transmission after a certain timeout time has passed. This is done, e.g., in the TCP

L. Gabrielli and S. Squartini, *Wireless Networked Music Performance*,
SpringerBriefs in Electrical and Computer Engineering,
DOI 10.1007/978-981-10-0335-6_3

transport-level protocol, or in wireless protocols it is done at lower levels when the received packet checksum is incorrect. However, such procedures require time and are not suitable for time-critical application such as NMP (unless transmission time is $\ll$ than the deadline imposed by audio buffering). Furthermore, in NMP the only viable solution is to trust the network robustness to failures and losses or to apply some redundancy, in order to greatly reduce the probability of lost packet. Under the hypothesis of a random distribution for packet loss, redundancy can be exploited. However, if the losses are correlated, e.g., due to a link prolonged failure, redundancy provides no improvement.

Packets may also arrive at the receiver after the time they should have been sent to the audio card for playback. This is discussed in Sect. 3.2. Moreover, when the clock rates of the transmitter and receiver audio cards are not synchronized, packet loss may occur as described later in Sect. 3.7.

Finally, one of the two machines, may fail in correctly scheduling the audio processes, so that the audio card is not timely supplied with or read for new audio data. Issues with process scheduling is briefly recalled in Chap. 5, where a practical implementation of audio capture and playback under a Linux operating system is described.

Unfortunately, even with a good NMP deployment and best networking conditions, it is a good practice to consider occurrence of dropouts. Loss concealment is a good practice to avoid all the dropouts due to the network and clock issues. For this reason many audio coding and decoding algorithms also include error or loss concealment. Indeed, when dropouts cannot be avoided, the last tool to resort to is psychoacoustic masking. Currently, the OPUS codec (formerly CELT), is widespread in audio streaming application for its very low delay (less than 10 ms), objective audio quality, and loss concealment [1]. In a recent paper, low-latency error concealment for NMP is reported [2].

Errors can be concealed too, and though communication technologies are nowadays very robust to errors, a packet may still get corrupted. Generally, packets containing errors (which are calculated through checksum or similar mechanisms) are discarded at the receiver end, however, some transport protocols such as *UDPlite* allow the user application to forward a corrupted packet for further error correction or concealment in software.

While planning a NMP deployment, it is a good practice to figure out an upper bound to dropouts during a session. A strict bound could be zero dropouts tolerated for an entire session. In practice, however, dropouts can be expected in exceptional conditions. The team behind the LOLA project (see Sect. 2.2.3), e.g., in a workshop presentation suggested a packet loss ratio smaller than 0.3 %.[1] In general, a constraint may be decided to be q dropouts per session duration s, i.e., a probability

$$p_t \leq q/s. \tag{3.1}$$

[1] In that context a packet loss was equivalent to a dropout.

In Sect. 5.7, results are reported for packet loss over a wireless link. In that case the PLR, i.e., Period Loss Rate is reported, that is the rate of audio buffers that are lost, which is more psychoacoustically meaningful than network packets. An adequate psychoacoustic research work could investigate further on the topic, to link PLR to perception of glitches from the audience. It is possible, e.g., that the loss of two consecutive or very near audio periods is perceived as one glitch. Furthermore, the audience may not expect dropouts at first, thus, cognitive effects may make the listener unconsciously discard a dropout.

3.2 Latency and Interaction

As reported above, a variable amount of latency is inherent to NMP and determines the feasibility of the performance and the approach to consider. The approach of most interest is the RIA, being the most technically challenging and requiring the lowest possible latency. Where not otherwise specified latency is defined as the delay it takes for the sound to reach the remote listener from the source, while the Round-Trip Time (RTT) will be referred to as the time the audio takes to travel from the source to the remote end and back to the source. The RTT is important in assessing the ability of a NMP setup to build an interactive interplay between musicians.

As reported above, Carôt takes the 25 ms figure as a threshold for latency in RIA, drawing from [3]. This threshold is accepted by several authors and is called Ensemble Performance Threshold (EPT). There are, however, other studies yielding different values and points of view. First and most importantly, most studies conclude that the tempo change slope can be fitted with a linear model to a good approximation, thus, with increasing delays the tempo decreases [4, 5]. In [6] a simple model for human performer tempo deceleration based on perfect memoryless tempo detection is first assumed and then proved inaccurate in predicting human performers in an NMP context. The model assumes that tempo M decreases at each round according to

$$M(n) = \frac{60}{c_0 + nd} \tag{3.2}$$

where $c_0 = 60/M_0$, i.e., the quarter note interval in seconds with initial tempo M_0, n is the quarter note sequential number and d is the delay imposed by the network. A steady tempo would be kept, following this model only for $d = 0$. Tests proved humans to perform differently than the memoryless model. For this reason, Driessen et al. [7] investigated on human player modeling in the delayed performance scenario, following coupled oscillator theories, summarized in their paper. By fitting data a good model was found to approximate the human player behavior, i.e.,

$$M = M_0 - kM_0d \tag{3.3}$$

where the constant k is found to be $k \simeq 0.58$ from data fitting. The authors hypothesize that results obtained with simulated NMP setups are probably to be found in an acoustic environment where the subjects are situated at a sufficient distance to impose propagation delays similar to those of an NMP setup. The reader must not forget that given the sound propagation speed in STP conditions each meter adds approximately 3 ms delay. In an orchestra, for instance, a maximum of 46 ms is reached between the most distant players (hence the need for a conductor as a visual cue).

In [4, 8] it is found that at delays $\leq$11.5 ms the tempo accelerates. Farner et al. [5] report the same conclusions. Values are, however, different. In [5] musicians and nonmusicians are discovered to have different thresholds of acceleration, respectively, 15 ms and 23 ms. It must be noted, however, that the musicians were subject to a complimentary rhythm task (see Fig. 3.1) while the nonmusicians performed the same rhythmic pattern. It is unknown whether this difference was relevant to the results. Imprecision in timing was also evaluated in this study and delays over 25 ms were found to introduce imprecision. Acoustic conditions were relevant to imprecision: anechoic conditions imply a higher degree of imprecision, while room reverberant conditions imply a slightly lower tempo.

A last related study [9] reports results similar to the above, and states that until 60 ms the performance can endure harmoniously, breaking down only over that value.

All the aforementioned papers conducted tests at 90 BPM. Other tempos are worth investigating to obtain a more general model. Furthermore, they only investigate on tempo keeping tasks. Close to no reference is found in the literature regarding perceptual aspects related to the success of interaction, related to tempo, or the importance of latency for loose synchronized pieces, such as those found in the contemporary acoustic and electroacoustic classical music repertoire. There are even scores written with NMP in mind with little to no connection between the performers. Although latency constraints in these cases can be presumed to be less strict no specific finding is reported up to date.

One last factor to be assessed is the importance of video signals in NMP. There is discordance whether low-latency video signal is necessary for synchronization, whether they can improve reaction and interaction, or whether it does not greatly affect the performance. Since aligning audio and video requires audio to undergo the high latency imposed by video, it should be verified whether audio and video need to be perfectly aligned or some delay between the two can be tolerated.

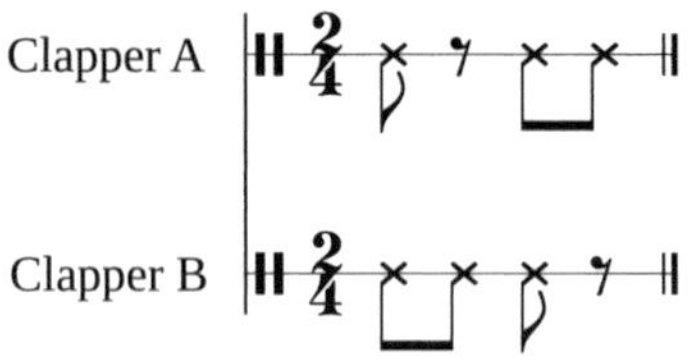

Fig. 3.1 Score of the clapping pattern used in Farner's and Chafe's complementary rhythm tests

Clapper A

Clapper B

3.3 Audio Blocking Delay

After this brief report on psychoacoustic findings related to latency, what are the factors that introduce latency in an audio link? With current audio equipment, one unavoidable source of latency is that related to audio buffering after or before the A/D and D/A conversion. The A/D and D/A converters are explicitly designed to convert audio in real time (the conversion time is totally negligible in this context), however, the issue is with the data storage and buffering. Since in most modern networking technologies data is exchanged in packets and computing architectures are more efficient with chunks of data—rather than on a sample-by-sample basis—the only feasible way to treat audio data in this field is to store and exchange buffered data. The buffer size and the sampling rate determine a blocking time, or a **buffer time**, which delays operations such as transmission. Thus, for a simple audio link, with one end acquiring and transmitting, and the other receiving and playing it back, a minimum latency is imposed by one buffer time at each end. The buffer is filled of several audio slots (at least two), called *periods*. Typical values for the period size with modern computing architectures are 128 to 512 samples at 44.1 to 96 kHz, i.e., 1.3 to 11.6 ms. The buffer time in modern computing architectures is not hardwired, but depends on a trade-off between CPU time and latency, since short buffer times lead to a higher interrupts frequency, and more operating system overhead. Embedded hardware for music employ shorter period sizes, e.g., 16, 32, or 48. For the sake of comparison, a digital mixer, typically imposes an input–output delay to the signal of 1.6 or 0.8 ms, while many digital keyboards reproduce sound at 32 to 44100 kHz and a period sizes of 16 to 64 samples. Figure 3.2 displays an A/D D/A system with two periods, each under exclusive control of the hardware or the software. These are exchanged each time they are filled by the A/D and emptied by the D/A. Timing is imposed by the hardware, which triggers interrupts after finishing memory transfer. At each interrupt the software loses control over the memory area occupied by the period it was writing/reading to, thus, it needs to write back the data to be played back before the interrupt to avoid audible glitches. The input/output latency imposed by the buffering mechanism is $2P$ with P being the period size. In one period time the software locally copies the period, processes it, and writes it to the output ring buffer. In other cases, one ring buffer is sufficient for both input and output if the hardware and the software can store temporarily the content. A more realistic diagram is reported in Fig. 3.3, where the software local copies and processes the period, and writes it back to the same memory area where it had read the current period. The hardware has FIFO buffer queues where samples computed by the A/D are shifted in and samples to be converted by the D/A are shifted out. The hardware must first read and then write samples to work correctly. This second case is equivalent to the first in terms of input/output latency.

Differently from the hardware, which is bound to a hardware timer (although, as later shown in Sect. 3.7 it is not as precise as it could be expected), the audio-related software applications, in general purpose Operating Systems (as those used in NMP contexts so far), share CPU time with other processes and are subject to scheduling

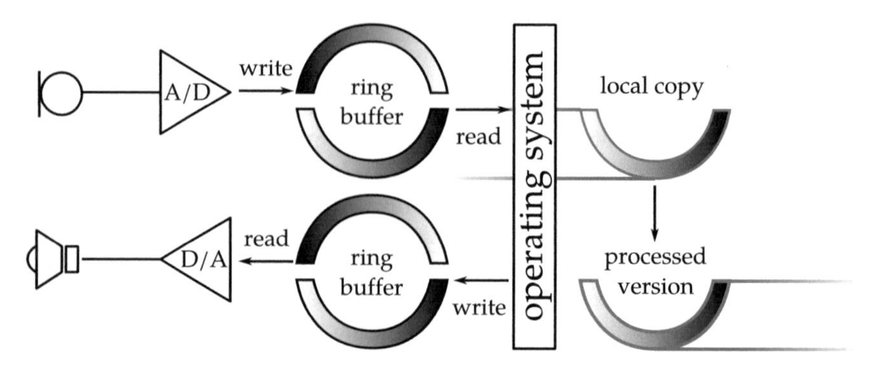

Fig. 3.2 Simplified graphic description of a A/D D/A buffering mechanism with ring buffering. In this case the periods employed for input or output are two (the minimum, also called *ping–pong* mechanism). Each period is either under exclusive control of the hardware or the software. The data transfer can be both in chunks or samplewise

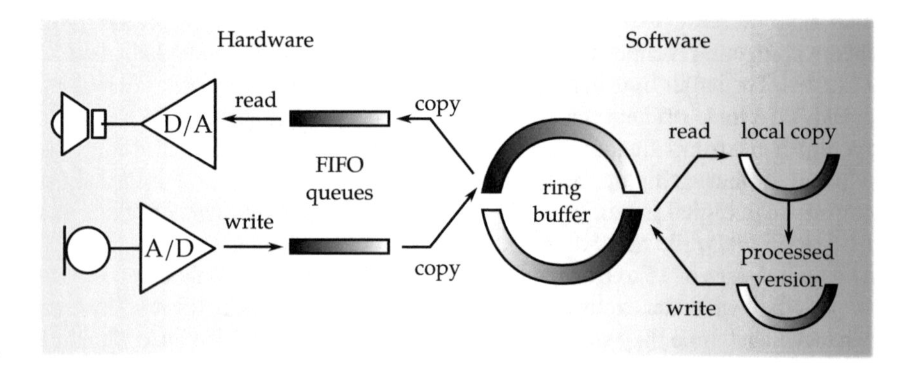

Fig. 3.3 A/D D/A buffering mechanism equivalent to Fig. 3.2 with one ring buffer mapped to memory and hardware FIFO queues to shift data in and out from and to the A/D D/A system

delays. Even if audio tasks can be scheduled with very high priority, they are subject to a certain delay after the audio hardware triggers an interrupt, depending on the software complexity and the number of nested function calls required to reach the section of code critical to audio buffering and audio processing. Furthermore, several audio tasks may compete for execution in the CPU. In single-core machines, the only possible solution is time sharing. In the Linux kernel, for instance, two alternatives are viable. Either a process takes exclusive control of the hardware, (nonetheless it may have several threading to compete for the CPU), either an audio server takes exclusive control of the hardware but provides mechanisms for other audio processing processes to read and write to the hardware. In the latter case, processes are scheduled based on the priority assigned to the audio thread and the dependencies between audio processes. If, e.g., audio process B is daisy chained to audio process A, it requires A to process audio first, pass its output to B and then it can execute. An audio server takes care of this prioritization, as it shall be detailed later. In multi-core processors, nowadays very common also in embedded devices, A and B can be parallelized if and only if they have no dependencies, e.g., if they both take input and write it back from and to the sound card. If one of the two, however, has not finished yet processing by the time the audio card reclaims the current memory area, an error condition occurs, which can be recovered in several ways, e.g., handing to the hardware an empty buffer.

To conclude, the application in charge of acquiring the signal and transmitting it to the network may be subject to a delay as large as the period between two interrupts minus its execution time. What is generally done for network packet transmission is relying on the operating system kernel for packet encapsulation and delivery. In general purpose operating systems, the packets are enqueued for later transmission to the hardware, according to some throughput optimization rule. This means that often the transmission is delayed. For this reason in LOLA [10] and in [11] network drivers have been rewritten in order to minimize this additional delay. At the receiving end similar considerations apply: the execution of the audio task that reads network packets is not instantaneous at packet reception.

3.4 Algorithmic Delay of Audio Compression

Once signals are discretized and quantized they need to be transmitted. When bandwidth is an issue, encoding and compression are employed. There is a vast corpus of textbooks, papers, and standards dealing with encoding and compression, however, in NMP two strong constraints must be taken into account:

- audio quality,
- algorithmic latency.

A total latency budget need be assessed for the specific performance. Performances requiring a real interaction and tight feedback (such as in classic, modern and jazz music) are subject to psychoacoustic findings of Sect. 3.2. Other performances such as contemporary classic, avantgarde, or those executions relying on free improvisation, drones, or a click track, are more tolerant to latency. As stated in Sect. 3.2, in the absence of quantitative results from research related to more loosely constrained executions, tests with the musicians should be undertaken well before the performance in order to assess the latency that can be tolerated. The network and input/output delays, discussed in this section, must be subtracted from the total budget to obtain the latency tolerable for the compression and encoding of the signal.

Several algorithms have been proposed for NMP up to date, with low latency and acceptable audio quality. Bandwidth and audio quality are generally two contrasting factors and a balance must be found empirically, if no other constraints are present. Latency also often trades for audio quality.

One reference in terms of low-latency performance is the ADPCM (Adaptive Differential Pulse Code Modulation). First developed at Bell Labs in the 1970s [12], it was introduced by Microsoft in the specifications for the WAV audio file format in the

1990s and is still supported by most operating systems. The compression method is simply based on encoding the differential information between consecutive samples with a lower number of bits with respect to the original sample. The quantization step is variable, and is often implemented as a variable gain before a fixed bit number quantizer. Although the algorithmic delay is very low (only two samples are needed for the difference), the resulting quality loss is often large. In [13], e.g., it is compared to another compression algorithm, later described, showing low scores in an objective PEAQ (Perceptual Evaluation of Audio Quality, ITU-R Recommendation BS.1387) comparison.

The most used compression algorithm for NMP is CELT [1], now part of the Opus standard [14], together with SILK [15]. CELT (Constrained Energy Lapped Transform) employs a modified discrete cosine transform (MDCT) with very short windowing. The algorithmic latency depends on the window size. Short window sizes yield a very low latency, but also a low spectral resolution. The MDCT spectrum is divided into 20 critical bands for psychoacoustic reasons, and the MDCT bins in each band are normalized, i.e.,

$$\mathbf{x}_b(i) = \frac{\mathbf{z}_b(i)}{\mathbf{z}_b^T(i)\mathbf{z}_b(i)} \tag{3.4}$$

where $\mathbf{x}_b(i)$ is the vector of the normalized coefficients and $\mathbf{z}_b(i)$ is the vector of the MDCT bins for frame i and band b. $\mathbf{y}(i)$ *innovation* coefficients are transmitted which are quantized versions of the difference between $\mathbf{x}(i)$ and past MDCT coefficients, normalized and scaled. A time-domain cross-correlation pitch extractor is employed, as in speech codecs, and takes into account past samples of the resynthesized signal (thus the encoder must also implement decoding). The use of CELT and, later, Opus, has been proposed in several papers for Networked Music Performance [16, 17] and is currently supported by Netjack (an audio networking backend for the JACK open-source software) and Soundjack (a software project by Alexander Carôt) [18]. The typical delay of the CELT algorithm is 8.7 ms with frame sizes of 5.8 ms and a look ahead of 2.9 ms at 44100 Hz. Typical bit rates are in the range 32–96 kbps, with the higher ones of quality comparable to the ones of MP3 codec (MPEG-1 Audio LayerIII, standard ISO/IEC 11172-3) at larger constant bit rates.

Other viable alternatives to Opus are ULD [19], also employed by Soundjack [20] and AAC-LD [21]. The latter has a latency of 34.8 ms at 44100 Hz, which is hardly recommendable for NMP. ULD, instead have a typical latency of 5.3 to 8 ms, very well suited to NMP. In principle ULD is a variable bit rate coding algorithm, based on time-domain operations, to avoid buffering delays typical of spectral domain algorithms. It employs an adaptive linear filter (*pre-filter*) which is driven by a perceptual model for *irrelevance reduction*. The perceptual model is based on DFT transform of the input signal. The size of the DFT transform windows is crucial for latency, since the input signal must be synchronized accordingly by delaying. For this reason window overlapping is used. Typically, a 256-bins DFT is calculated on a 256 window with 50 % overlap, introducing a 128 sample delay. The perceptual model estimates a masking threshold, transformed in filter coefficients for the pre-filter via Levinson–

Durbin recursion. Practically speaking, this operation is equivalent to normalizing (a variable gain is also applied) with respect to the masking threshold, for later quantization (rounding). Redundancy is reduced by means of predictive coding (see [22] for more details). Since this is based on backward prediction and entropy coding, it produces a variable bit rate output. For this reason in [19], the ULD scheme is loop controlled to maintain a constant bit rate. In very simplistic terms, a bit counter observes the bit produced at the output of the encoder and automatically controls a second variable gain applied before the quantization step. This implies a variable quantization noise.

Of high interest for its open-source nature is the *Wavpack* codec.[2] This codec is released under BSD license and is completely free software, employing only well-known algorithms and explicitly avoiding patented signal processing techniques. This codec supports lossless and lossy modes. The latter can achieve constant bit rates, necessary for NMP applications, where bandwidth must be reserved or not exceeded. As already stated, in NMP applications audio must be processed in small blocks. Since the lossy mode loses compression efficiency with small audio blocks, in [13] it has been extended with an additional layer that removes redundant data and hence achieves a good compression efficiency in this case as well.

Of all the aforementioned compression techniques, several can be taken into consideration for NMP, depending on the constraints, the technical challenges and aesthetic consideration to be taken into consideration for the singular performance at hand. Lastly, uncompressed audio streaming should be considered if the bandwidth allows it, to remove encoding latency.

3.5 Network Delay

In networked music performance, the most obvious factor of latency is the network delay. Two main factors affect network delay: the physical component to delay due to propagation of the signal being transmitted through the medium and the component due to medium access and packet switching. The first issue is of primary importance in the case of remote NMP where the network is a dedicated link, or a link with very simple routing and dedicated bandwidth. Of this nature are, for instance, the fiber-optic links that connect research institutions, such as the Internet2 in North America or GEANT in Europe. With these network it is possible to reserve bandwidth and program routing for the performance in order to minimize packet switching and routing path. Depending on the medium the signal propagation can be somewhere around 0.7 times the speed of light in void, i.e., about 5 ms every 1000 km. Unfortunately, the routing between distant locations can be much longer than the direct path.

In all other conditions the packet switching delay is in proportion to the greatest component to latency. In the case of general xDSL (Digital Subscriber Line) contracts, no bandwidth reservation, Quality of Service (QoS) protocol, or warranty in

[2]http://www.wavpack.com.

latency can be assumed, thus the path, even if only a few kilometers long can take as long as several tens of milliseconds. The Internet is also very variable in the delay it delivers, given the extremely fast time variability of the traffic it hosts. The Internet, thus, introduces a large packet delay variation (PDV, or *Network Jitter*), which is unpredictable. A DSL link also introduces overhead and other issues that must be considered [23].

Local Area Networks (LAN) represent the best scenario for NMP in terms of propagation delay, given the short distances the signal has to traverse. Unfortunately, this does not mean that network delay can be neglected. Switching and medium access are very dependent on the kind of network infrastructure employed. Current 100BASE-TX Ethernet technologies allow data rates up to 100 Mbps over Cat5 cables, or 1Gbps for 1000BASE-T Gigabit Ethernet. In now obsoleted shared media Ethernet networks, e.g., 10BASE-5 and 10BASE-2, a procedure for medium access and collision detection, Carrier Sense Multiple Access With Collision Detection (CSMA/CD) was employed. This included sensing the medium and transmitting if it was found to be idle. Collisions could still occur, and they were detected physically. In the case of a collision, random *backoff* times were to be waited by each colliding transmitter for the next retransmission, minimizing the chance of error, but introducing a factor of unpredictability to the transmission timing. Current Ethernet networks are full duplex and each path is separated from the others by switches, avoiding collisions. This ensures a reduced variance in the expected transmission time. As a best case for practical evaluation of an NMP prototype system, a network composed of two or more ends, connected via Gigabit Ethernet through a capable switch can be expected to be the most economic while effective means for a low delay, low network jitter connection.

Wireless LANs on the other side are totally different. We shall consider IEEE 802.11 networks, for reasons later detailed in Chap. 4. This family of standards, first released in 1997, expanded in the years to increase throughput capabilities, ease network and connection managing, improve security, etc. The main protocols for what concerns the physical level, the data rate and the medium access are 802.11a, 802.11b, 802.11g, 802.11n, 802.11ac, 802.11e. Two networking topologies are generally associated with IEEE 802.11: Ad Hoc, i.e., point to point and the star topology, that requires an *Access Point* (AP) to mediate among several *Stations*. Other topologies, e.g., mesh are defined but rarely implemented in software drivers or supported by commercial applications. In 802.11 a basic configuration of one AP and stations participating to a network is called a Basic Service Set (BSS). Although in amendment 802.11s a BSS comprised only of stations connected in mesh fashion is described, let us consider only the star topology BSS. All the nodes participating to a BSS operate in a predefined frequency band. Time is shared between the nodes in the following manner: a Superframe, announced by a Beacon Frame, is a periodic interval in which two different intervals take place, a contention period (CP) and contention-free period (CFP) (see Fig. 3.4). In CP all the stations willing to transmit compete to access the medium, while in CFP transmissions are scheduled by the AP after polling the stations for possible data to transmit. The medium access methods for the two periods are, respectively, Distributed Coordination Function (DCF) and

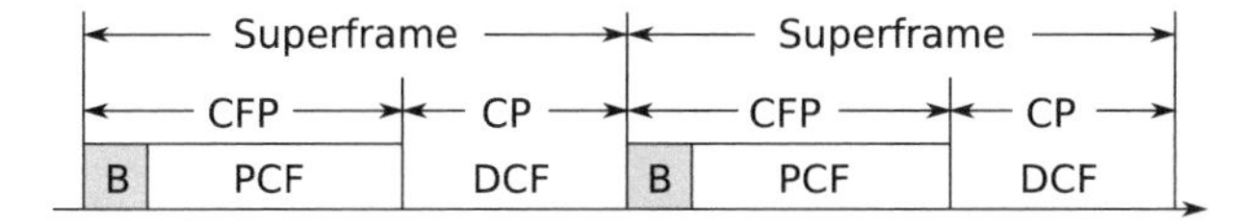

Fig. 3.4 In IEEE 802.11 time can be divided in superframes, each composed of a contention-free period and a contention period

Point Coordination Function (PCF). The latter allows for time-bounded traffic to be delivered within the CFP. Transmissions in PCF are interleaved by a PCF Interframe Space (PIFS). The DCF, instead, employs a Carrier Sense Multiple Access with Collision Avoidance (CSMA/CA). With this mechanism, a node willing to transmit senses the medium for a specified time frame (called DIFS, or DCF Interframe Space). If found busy it must wait for the medium to be idle. When this condition occurs a random *backoff* time is waited, to avoid collision with other nodes following the same procedure.

An issue with wireless networks is the hidden node problem: let us consider three nodes, with nodes A and C unaware of each other for the large distance and node B in between and in reach with both. If both node A and C want to transmit to node B at the same time they would incur into collision without knowing because they cannot sense each other. To partially overcome this problem IEEE 802.11 provides an optional mechanism of Request to Send and Clear to Send (RTS and CTS), that imposes to the nodes A and C to send a RTS prior to transmit, and node B to send a CTS to one of them to allow only one of the stations to transmit. This does not solve the problem completely and collision may still occur. An acknowledge (ACK) is sent by the receiver if no collision occurred, otherwise retransmission can be performed. A Short Interframe Space (SIFS) time must be waited before sending a CTS or an ACK.

In terms of effective throughput all the aforementioned mechanisms imply a large overhead. All the Interframe Spaces (PIFS, DIFS, SIFS, and others) imply that the medium stays idle for a certain time. Consider that the three aforementioned Interframe Spaces have durations of tens of microseconds and the superframe can have durations of the order of 1 ms. Depending on the data rate, the traffic, and other parameters, the idle time of the wireless medium can be just one order of magnitude below the superframe, yielding a loss of several percent point of the throughput. The backoff time, possible collisions, and retransmissions must be also considered that reduces throughput consistently.

In terms of latency the network is far from being ideal, too. The delay is not predictable, due to backoff times, number of stations, and so on. Time-bounded transmission as those required for NMP could be scheduled with PCF. Time slotting can be of high benefit to reduce overheads and reduce variance of the network delay. Unfortunately, it turns out that most commercial 802.11 chipsets, equipment, and hardware drivers do not provide PCF, nor they implement a more recent variant proposed in 802.11e named Hybrid Coordination Function Controlled Channel Access (HCCA) [24].

To summarize salient figures affecting the network delay and its variance for NMP usage are:

- transmission in contention or CF periods,
- periodicity of the superframe,
- number of stations in the BSS,
- Maximum Transmission Unit time (MTU) and bandwidth needed by each audio transmitter.

The reader who is already aware of the basics of IEEE 802.11 may object that recent 802.11 implementations also introduce Quality of Service (QoS). Although this is correct, in audio networking QoS is hardly useful. Indeed, QoS is conceived to impose priorities to different kinds of payload. In IEEE 802.11e, a new access mechanism to provide QoS and optionally replace DCF is proposed, named Enhanced Distributed Coordination Access (EDCA), which defines eight levels of priority and four categories of payload: background, best effort, video, and voice, with increasing priority. Defining traffic categories in a general-purpose wireless network allows time-bounded packets to reach destination in shorter time with respect to other kind of traffic. It is the case, for instance, of a small office network, where traffic from a video conferencing call need to be prioritized with respect to traffic related to a document to printed or simple network synchronization and update (e.g., device discovery). In a wireless network setup for NMP, however, most of traffic is related with audio or control (e.g., MIDI) which need to have low latency, and possibly very predictable. There is little chance, thus, to impose priorities. The only traffic that can be assigned a low priority is, thus, data logging, debug information, device discovery, and session control data.

3.6 Delay Jitter

The network delay is not constant, and can quickly vary from packet to packet. Network delay is said to be, thus, subject to jitter. In informal terms, a buffering system is needed at the receiver to compensate for this jitter, i.e., the receiver must store the received packets, possibly reorder them, and wait for late ones before sending them to the hardware for playback. The larger the buffer, the safer the playback and the larger the latency. For this reason, a trade-off must be found, depending on the kind of performance and several other factors. Characterizing the typical network jitter is useful to properly dimension the buffer. A brief formalism for network delay jitter is, thus, provided. Abstracting from the communication medium and network topology, a certain delay will occur between the transmission of a packet and its reception by the recipient. This delay is subject to jitter, i.e.,

$$n(i) = \bar{n} + \nu(i), \tag{3.5}$$

where the time-varying delay $n(i)$ depends on a constant component $\bar{n}$ (propagation speed, switching, etc.) and a stochastic process $\nu(i)$. The stochastic component cannot be predicted, thus a buffering mechanism must be employed at the receiver side by allocating a buffer of *NP* samples. Choice of the buffer size depends on the shape of the probability density function (PDF) estimated for $\nu(t)$. If the PDF $\nu(\tau)$ reaches zero at ∞, a probability—however, small—of dropout must be tolerated. A probability of tolerated dropouts p_t in the form of Eq. 3.1 must be chosen. In the case, e.g., of one dropout tolerated over a 1 hour performance

$$p_t = \frac{1}{\frac{F_s}{P} \cdot 3600} \tag{3.6}$$

with F_s being the sampling frequency, P the fixed packet payload size in terms of audio samples. Once the PDF is known, the maximum delay D_M is the delay value for which the residual area (the tail of the PDF) is $A(D_M) \leq p_t$, as clarified by Fig. 3.5. The buffer size should be $NP \geq D_M$ to accommodate for late packet arrivals up to the maximum delay. For each specific application, the PDF should be characterized. This, however, is often difficult, and the PDF may be highly time-varying. It also does not take into account other impacting factors such as momentary link failure due to accidents, network failure, power shortages, and such. Thus, more pragmatic strategies for buffer size choice may be employed.

To summarize, network jitter cannot be avoided, but mitigated or compensated for, by estimating a worst case and allocating sufficient buffering to allow late packet arrival. However, the worst case condition may generate a buffering latency too high to sustain NMP, therefore a trade-off must be done. In that case a late packet may generate a dropout, i.e., arriving too late for the software to send the data to the audio hardware. A few graphical representation of the concept described above are proposed. Figure 3.6 reports the minimum number of delays to take into account for an audio link:

- the audio blocking time equal to one period to be recorded (orange),
- the delay involved with calling an interrupt service routine, waiting for the audio thread to be scheduled by the operating system and the possible algorithmic delay for encoding and compressing the audio,

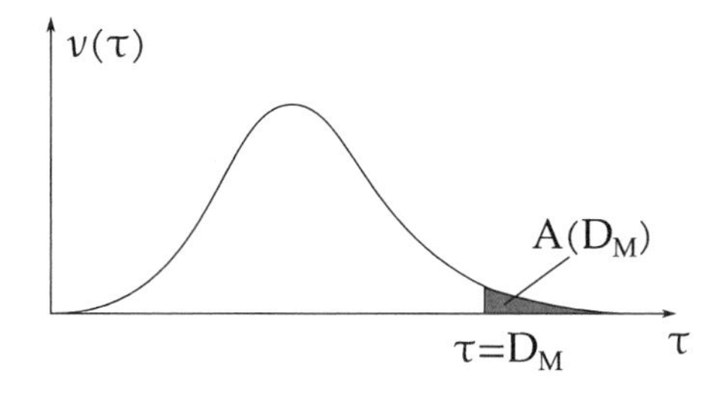

Fig. 3.5 An example probability distribution function of the jitter and choice of the buffer size accounting for the maximum tolerated delay D_M

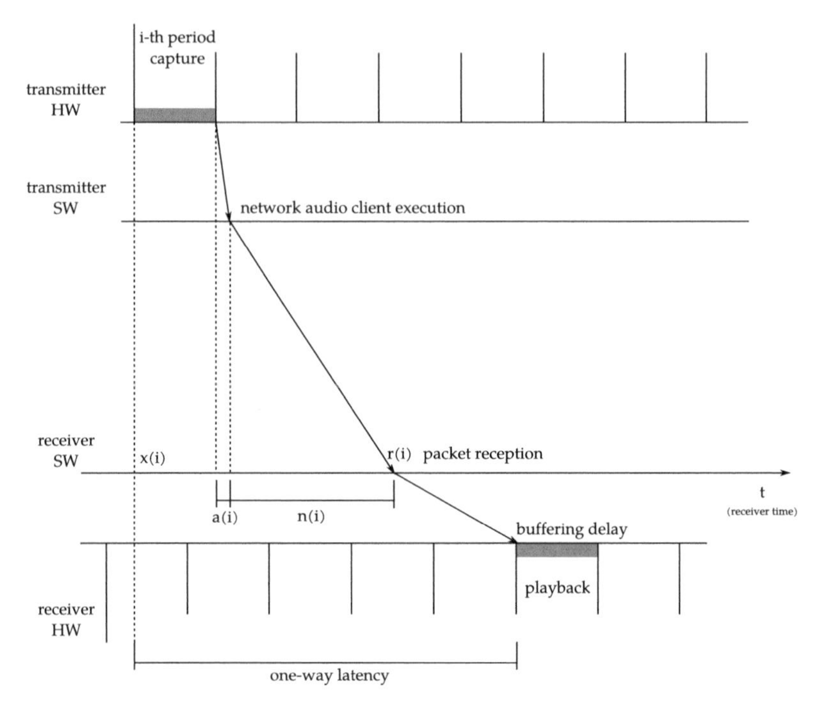

Fig. 3.6 Delays related to an audio link. Interrupt servicing delay and activation of the network audio thread, network delay, and queuing delay

- the time the packet spends in the operating system queue for packets to be transmitted,
- the delay to cross the network and reach the receiver,
- the algorithmic delay for decompression and decoding,
- the buffering the receiver requires to take into account network jitter and reordering,
- the audio blocking time equal to one period to be played back.

The network jitter effect is totally removed by the receiver buffering mechanism when $n(i) < NP$, otherwise a dropout occurs. In Fig. 3.7, e.g., a packet affected by a random delay and arriving in the range $t_i < t < t_{IRQ2} - \varepsilon$ (with ε very small) is played back in the same audio card time slot. Provided that the receiver incorporates a packet reordering mechanism and the packets are provided with a counter by the transmitter, a packet arriving between T_i and T_{IRQ2} can be played back properly. Furthermore, latency is insensitive of the exact arrival time, due to quantization of time in periods.

In Fig. 3.8 an example transmission where the latency is initially set by the transmission delay imposed by packet 1, incur into dropout of packet 4 when the network delay for that packet exceeds the buffering capacity of the receiver. A larger buffer would take into account for additional jitter in the network delay and, accordingly, allow packet 4 to be scheduled for playback notwithstanding its large delay.

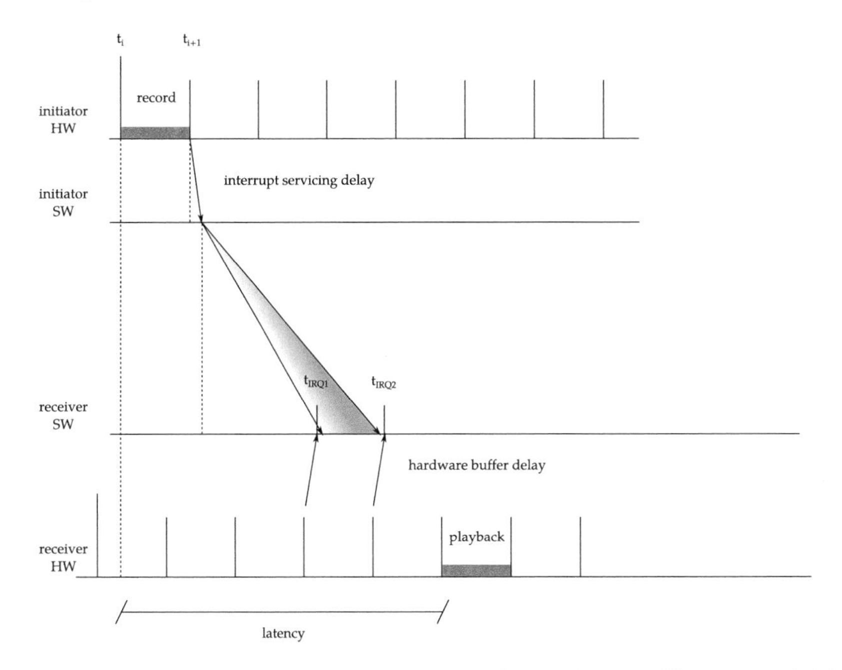

Fig. 3.7 A packet arriving slightly after IRQ1 or right before IRQ2 due to different network delay has the same effect on latency due to quantization of time in periods. The larger the period (and the buffering) the more tolerant the system to network jitter

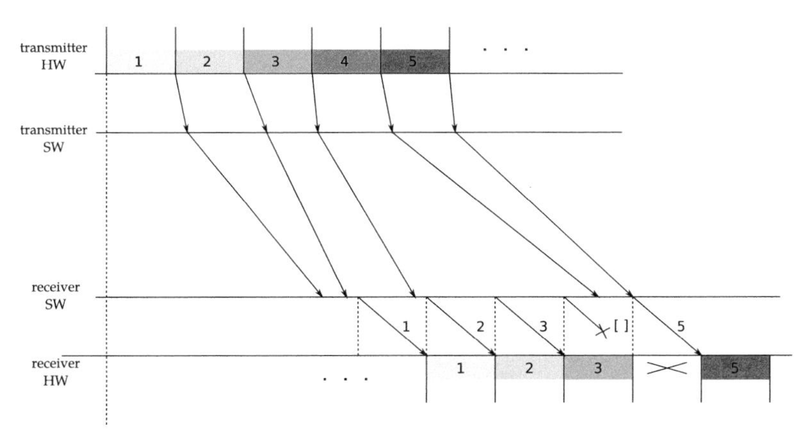

Fig. 3.8 An example of packet dropout due to network delay jitter. The larger delay imposed by network to packet 4 forbids it to be played back as it arrives too late. The average latency is imposed by the first packet. After being received all the packets are issued to the hardware buffer and are, thus, delayed by the hardware buffer blocking time. For this reason, packet 4, even if arrives before the actual time it is played, cannot be issued to the hardware anymore

3.7 A/D and D/A Clock Synchronization

Another technical issue faced in NMP is the nonideal behavior of the audio card hardware clock. Two hardware clocks, notwithstanding their precision, can drift with respect to each other. For this reason several digital audio networking technologies transmit a master clock that is received and reconstructed through a Phase-Locked Loop (PLL) by the receiver, similarly several inter-chip communication bus or digital communication technologies. The PLL reconstructs the clock and reduces phase differences introduced by the transmission medium or noise to a great extent. The master clock can be transmitted through a separate channel, as it is done with the so-called word clock with some professional audio gear, using coaxial cables. In other cases, the clock is implicitly embedded into the signal emitted by the transmitter and can be reconstructed by the receiver by locking the PLL to a preamble or a part of a message., such as AES3 or S/PDIF the two apparati share the same medium. The transmitter, thus, injects digital samples at a certain audio rate and Differently from several audio networking technologies that share the physical medium and extract the Even on an ideal network, there is another issue that is worth addressing which is audio card clock skew, i.e., a slight difference between the nominal clock and the actual value. This may be due to errors in components rating as well as a change in environment conditions such as temperature. Even a small difference between the two remote ends clocks can lead to dropouts in a relatively short time frame. For instance, a 20–60 ppm skew (e.g., 1–3 Hz at 48 kHz) with short buffers can lead to dropout at the receiving end in a few minutes. Assuming a buffering mechanism at the receiving end compensating for network jitter, of N slots of B samples each, and assuming (for simplicity) that at both ends the machines process audio buffers of B samples with the same nominal samplerate but different actual clock frequencies F_1 and F_2, a dropout will occur every

$$\Delta = \frac{N \cdot B}{|F_1 - F_2|}. \tag{3.7}$$

For a skew of 1 Hz at 48 kHz nominal samplerate with buffers of 128 samples and a circular buffer of 4 slots this yields to 8.5 min, i.e., 7 dropouts in 1 h, which is not acceptable for a regular performance. Depending on the buffering mechanism the Δ may be even lower. Let us consider a circular implementation of the buffer, i.e., a ring buffer. Without prior knowledge regarding the two clocks frequency the safest way to initialize the read and write pointers is at the two furthest position, i.e., one at the slot $i = 1$ and the other at slot $j = N/2 + 1$ (assuming an even N). The underrun condition occurs when the read pointer proceeds faster than the write pointer (i.e., the remote end has a slower clock) and reaches the write pointer slot. The overrun condition occurs when the write pointer proceeds faster than the read pointer and finally has no empty slots to write to. With the aforementioned initial conditions i, j, the time between two consecutive overruns Δ_o or underruns Δ_u is $\Delta_o = \Delta_u = \frac{1}{2}\Delta$.

To recover from underrun state, the read pointer is spun back m slots, thus it reads again part of the previous audio (or simply null data if these were erased after reading). To recover from overrun, recovery is done by advancing the write pointer as it is supposed to do and place the read pointer m slots ahead. With such a ring buffer implementation, if

- $F_1 > F_2$: $\Delta_o = \frac{m}{N}\Delta$,
- $F_1 < F_2$: $\Delta_u = \frac{m}{N}\Delta$,

Clearly, m can be at best $m < N - 1$. The closer to 1 the ratio m/N is, the closer to the ideal case.

To express the problem in more rigorous terms the following formalism is introduced. Any hardware clock source is employed to obtain a system time that is used in software for many purposes. A clock can be seen as an oscillator generating periodical events and a (software) accumulator that increases its count at each event. An accumulator function, or time function $T(t)$ [25], is a mapping between the real time t and the clock events, i.e., a piecewise continuous function that is twice differentiable

$$T : \Re \longrightarrow \Re. \tag{3.8}$$

Let $T_1(t)$ and $T_2(t)$ be two independent clock generators, and $T_1'(t)$, $T_2(t)'$ their time derivatives, then the:

- **offset**: is defined as the difference $\rho_{21} = T_2(t) - T_1(t)$;
- **frequency**: is the rate at which a clock progresses. The instantaneous frequency of the first clock at time t is $F_1(t) = T_1'(t)$;
- **absolute skew**: is the difference between a clock frequency and the real time, e.g., for the first clock $\beta_1 = T_1'(t) - t'$;
- **relative skew**: is the difference between two clock frequencies, e.g., $\sigma_{21} = T_2'(t) - T_1'(t)$;
- **clock ratio**: the ratio between two clock instantaneous frequencies, e.g., $\alpha_{21} = T_2'(t)/T_1'(t)$.

When a sentence clearly refers to the frequency difference between two clocks, the relative skew is simply called skew.

The general model for a clock time function is

$$T(t) = \beta(t) \cdot t + \xi(t) + T(0), \tag{3.9}$$

where β is the absolute skew, which determines the slope of the function, $\xi(t)$ is a power-law random process modeling jitter of the clock events, and $T(0)$ is the initial value of the function. With constant $F_1 \cdot F_2$, the relation between clock ratio and skew is

$$\sigma_{21} = F_2(t) - F_1(t) = \alpha F_2 - F_1 = (\alpha - 1)F_1 \tag{3.10}$$

In nonideal oscillators, unfortunately, the frequency $F(t)$ is slowly time-varying and departs from the nominal value F_0, i.e.,

$$F(t) = F_0 + F_e + \nu(t), \tag{3.11}$$

where F_0 is the nominal frequency, F_e is a frequency offset, i.e., a constant departure from the nominal value and $\nu(t)$ is a slowly time-varying stochastic process.

Instantaneous synchrony takes place when $T_1(t) = T_2(t)$, meaning that time stamps generated by two different machines at a specific time t are the same. Of greater interest for a long run synchronization of two remote ends is the case when $T_1'(t) = T_2'(t)$, i.e., two clocks have same pace, and $\beta_1 - \beta_2 = 0$. Otherwise, the offset between the two time functions diverges $T_2(t) - T_1(t) \rightarrow \infty$ for $t \rightarrow \infty$. Clock frequency synchronization is also called along the essay *relative time approach*, while *perfect* or *absolute synchronization* is the case when both $T_1(t) = T_2(t)$ and $T_1'(t) = T_2'(t)$.

Several synchronization mechanisms are proposed for wireless networks. In [26], e.g., a mutual synchronization mechanism between two or more ends is proposed employing a control loop that minimizes the error between two time functions. This mechanism, called CS-MNS, is an absolute synchronization mechanism. In the audio case, however, the time function generally starts when the audio process is started. Relative time synchronization is already sufficient since two time functions may have different offset if two audio processes are started independently. A last approach worth to mention here was proposed by Carôt [27] that employs an external frequency generator driven by software. The frequency generator replaces the sound card clock. Unfortunately, the approach requires a frequency generator and a sound card that is capable of receiving a so-called word clock, i.e., an external clock reference. Two approaches for synchronization through resampling are provided in the next chapter.

It must be noted that professional equipment is prone to clock skew as well as inexpensive OEM hardware.

3.8 Audio and Video Synchronization

From the NMP survey conducted in the previous chapter, no established practice emerges for audio and video transmission synchronization. Some technical solutions do not even attempt at synchronizing the two heterogeneous streams, leaving the video to lag up to 1 s behind audio. Early implementations employed proprietary videoconferencing solutions that imposed a large latency and an artificial delay was added to the audio to synchronize with the video system [28]. The same team (McGill Ultra-videoconferencing Research Group), later considered very low latency audio and video transmission and developed a software system to allow uncompressed and compressed audio and video transmission. The system allows both unidirectional and bidirectional audio and/or video connections, with and without compression. Audio and video are transmitted separately and on different UDP ports, and no effort is done to synchronize the two streams. Synchronization can be obtained only manually by an artificial delay and the dropout tolerance (i.e., buffering) can be adjusted independently for audio and video. With DV streams (conforming to IEC

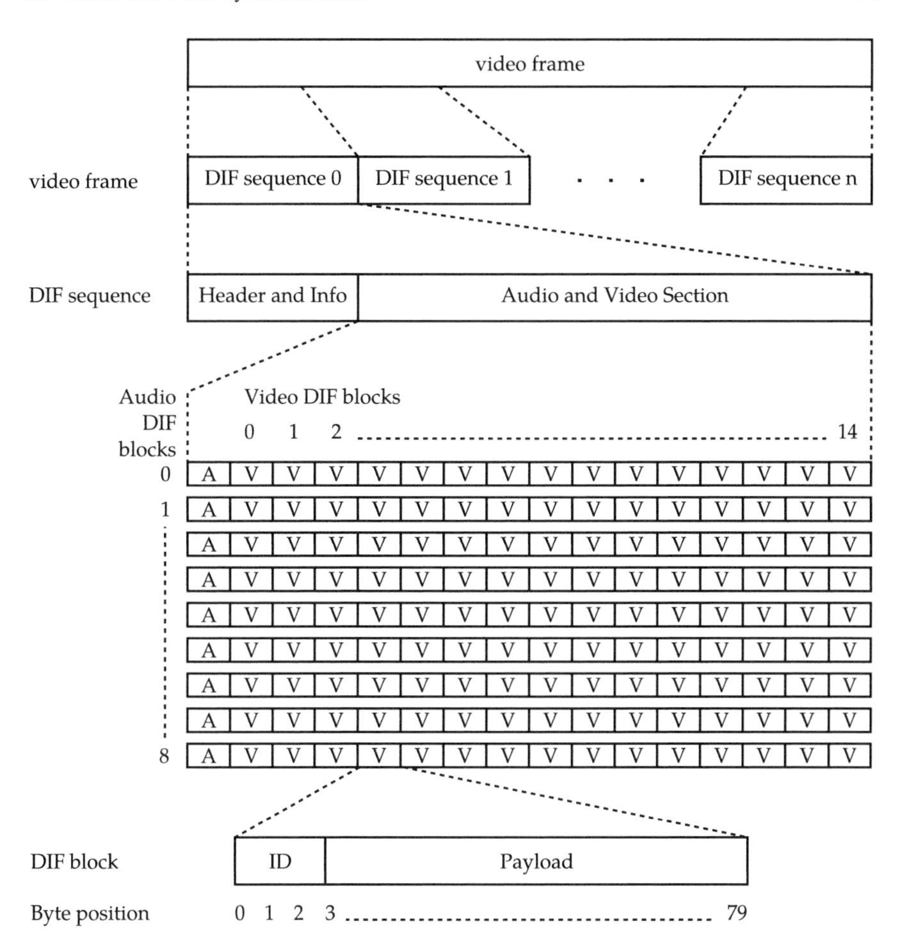

Fig. 3.9 A simplified representation of the DV DIF sequence audio (DIF blocks labeled "A") and video interleaving (DIF blocks labeled "V")

61834-2), however, which have audio and video bundled together, data is transmitted as a unique stream. It is worth mentioning that in the DV format [29], and similarly in other related standards that stem from it (e.g., SMTPE 314M-1999), the audio data is interleaved with video data and generally the audio clock is locked to the video clock. Figure 3.9 visually reports the composition of a video frame, highlighting the interleaving of audio and video data.

The Ultragrid software, introduced in Sect. 2.2.3, employs similar strategies for audio and video synchronization. If the source is a single device with audio and video capture which streams the audio data embedded with video in HDMI (implementing ANSI/CEA-861-F standard) or as *ancillary data* in HD-SDI format[3] then the stream is transmitted as a whole. In all other cases, no audio and video synchronization is

[3]SMPTE 291 describes the ancillary data and SMPTE 292 the video data format.

enforced by the software, with audio transmitted at very low latency. In Ultragrid compressed video can reach a minimum of 83 ms of latency, given by the need for 5 consecutive video frames for compression at 60 fps. Audio and video are streamed to different RTP ports, with two additional RTCP ports (RTP Control Protocol) to manage the transmission of the streams.

The LOLA project software supports a set of audio and video capture devices but no interleaved audio and video. It also does not synchronize or interleave the two data streams, but transmit as soon as it receives useful data to two different UDP ports. The packet size is fixed to around 1 KB for both audio and video streams. Since the audio period size is fixed to 32 or 64 samples and audio can have an arbitrary number of channels up to 10 (64 samples) or 20 (32 samples), the actual payload to send may largely vary.

The Soundjack software, was started in 2005 as an audio-only NMP software by Carôt. Nowadays, it is still active and in a paper in 2011 a video streaming transmission was proposed [18] with audio and video interleaved, by reducing video in small chunks to be sent in the same packet with audio. In 2014 the author claimed[4] that the first video implementation was left and was reimplemented in a Windows-only branch of the software.

References

1. Valin J-M, Terriberry TB, Montgomery C, Maxwell G (2010) A high-quality speech and audio codec with less than 10-ms delay. IEEE Trans Audio, Speech, Lang Process 18(1):58–67
2. Fink M, Zölzer U (2014) Low-delay error concealment with low computational overhead for audio over IP applications. In: Digital audio effects conference (DAFx-14), Erlangen, Germany, Sept 2014
3. Schuett N (2002) The effects of latency on ensemble performance
4. Chafe C, Gurevich M (2004) Network time delay and ensemble accuracy: effects of latency, asymmetry. In: 117th audio engineering society convention. Audio Engineering Society
5. Farner S, Solvang A, Sbo A, Svensson PU (2009) Ensemble hand-clapping experiments under the influence of delay and various acoustic environments. J Audio Eng Soc 57(12):1028–1041
6. Gurevich M, Chafe C, Leslie G, Tyan S (2004) Simulation of networked ensemble performance with varying time delays: characterization of ensemble accuracy. In: Proceedings of the 2004 international computer music conference, Miami, USA
7. Driessen PF, Darcie TE, Pillay B (2011) The effects of network delay on tempo in musical performance. Comput Music J 35(1):76–89
8. Chafe C, Caceres J-P, Gurevich M (2010) Effect of temporal separation on synchronization in rhythmic performance. Perception 39(7):982
9. Kobayashi Y, Miyake Y (2003) Analysis of network ensemble between humans with time lag. In: SICE 2003 annual conference, Aug 2003, vol 1, pp 1069–1074
10. Drioli C, Allocchio C (2012) LOLA: a low-latency high quality A/V streaming system for networked performance and interaction. In: Colloqui informatica musicale, Trieste
11. Baltas G, Xylomenos G (2014) Ultra low delay switching for networked music performance. In: The 5th international conference on information, intelligence, systems and applications, IISA 2014, July 2014, pp 70–74

[4] In a mailing list.

12. Cummiskey P, Jayant N, Flanagan J (1973) Adaptive quantization in differential PCM coding of speech. Bell Syst Tech J 52(7):1105–1118
13. Kurtisi Z, Wolf L (2008) Using wavpack for real-time audio coding in interactive applications. In: 2008 IEEE international conference on multimedia and expo, June 2008, pp 1381–1384
14. Valin J, Vos K, Terriberry T (2012) RFC 6716: definition of the opus audiocodec. Internet engineering task force (IETF) standard
15. Vos K, Jensen S, Soerensen K (2010) Silk speech codec. Internet engineering task force (IETF) standard
16. Valin J-M, Maxwell G, Terriberry TB, Vos K (2013) High-quality, low-delay music coding in the opus codec. In: Audio engineering society convention 135. Audio Engineering Society
17. Meier F, Fink M, Zölzer U (2014) The Jamberry-a stand-alone device for networked music performance based on the Raspberry Pi. In: Linux audio conference, Karlsruhe
18. Carôt A, Schuller G (2011) Applying video to low delayed audio streams in bandwidth limited networks. In: Audio engineering society conference: 44th international conference: audio networking. Audio Engineering Society
19. Krämer U, Schuller G, Wabnik S, Klier J, Hirschfeld J (2004) Ultra low delay audio coding with constant bit rate. In: Audio engineering society convention 117. Audio Engineering Society
20. Krämer U, Jens H, Schuller G, Wabnik S, Carôt A, Werner C (2007) Network music performance with ultra-low-delay audio coding under unreliable network conditions. In: Audio engineering society convention 123. Audio Engineering Society
21. ISO/IEC 14496-3:2001 (2001) Information technology—coding of audio-visual objects—Part 3: audio, ISO/IEC
22. Schuller G, Hanna A (2002) Low delay audio compression using predictive coding. In: 2002 IEEE international conference on acoustics, speech, and signal processing (ICASSP), vol 2. IEEE, pp II–1853
23. Carôt A, Krämer U, Schuller G (2006) Network music performance (NMP) in narrow band networks. In: Audio engineering society convention 120. Audio Engineering Society
24. Burton M (2009) 802.11 arbitration—white paper
25. Moon S, Skelly P, Towsley D (1999) Estimation and removal of clock skew from network delay measurements. In: INFOCOM '99. Eighteenth annual joint conference of the IEEE computer and communications societies. Proceedings. IEEE, Mar 1999, vol 1, pp 227–234
26. Rentel C, Kunz T (2008) A mutual network synchronization method for wireless ad hoc and sensor networks. IEEE Trans Mobile Comput 7(5):633–646
27. Carôt A, Werner C (2009) External latency-optimized soundcard synchronization for applications in wide-area networks. In: AES 14th regional convention, Tokio, Japan, vol 7, p 10
28. Cooperstock JR, Spackman SP (2001) The recording studio that spanned a continent. In: First international conference on Web delivering of music. Proceedings. IEEE, pp 161–167
29. IEC 61834-2:1998 (1998) Recording Helical-scan digital video cassette recording system using 6,35 mm magnetic tape for consumer use—Part 2: SD format for 525-60 and 625-50 systems, International Electrotechnical Commission

Chapter 4
Wireless Communication Standards for Multimedia Applications

Abstract Several wired local area networking protocols are being exploited by audio-over-ethernet or audio-over-IP protocols. They are reliable in sustaining audio communication, distribution, and retrieval and can serve other challenging scenarios such as distributed audio computing [1, 2]. On the other hand, wireless technologies are relegated, at the moment, as an auxiliary feature for remote control of digital devices (mixers, loudspeakers, etc.) neglecting the potential that wireless networking can have even in the mission critical field of live music performance. At the time of writing, the use of wireless communication for audio transmission in musical applications is limited to analog or digital point-to-point unidirectional links transmitting microphone or guitar signals to a signal mixer, to allow speakers or musicians for a higher freedom of movement, during live events. These technologies focus on robustness of the link and durability of the power source, but have several limitations in terms of flexibility: only point-to-point signal transmission is allowed, no networking can be performed and different devices communicate on different channels. Interoperability is also an issue. It is worth, thus, investigating whether there are wireless communication technologies to exploit to lay foundation for a wireless audio networking system able to guarantee appropriate bandwidth, medium access strategies, and range in NMP contexts. For the sake of completeness, a last section of this chapter is dedicated to academic literature dealing with wireless networking for transmission of control and session data only.

Keywords Wireless audio networking · Bluetooth · IEEE 802.11 · Millimiter wave bands

4.1 Bluetooth

In multimedia setups, more complex protocols are employed. The two most adopted solutions are encapsulating audio/video streams over IP or employing A2DP (Advanced Audio Distribution Profile) Bluetooth for audio-only links. Bluetooth has been explored in a few works for real-time audio transmission. Besides the A2DP profile, academic works attempted to overcome some technical limitations by

L. Gabrielli and S. Squartini, *Wireless Networked Music Performance*,
SpringerBriefs in Electrical and Computer Engineering,
DOI 10.1007/978-981-10-0335-6_4

introducing novel features. One reason why Bluetooth appealed to many researchers is its inherent time-slotting transmission mechanism which is suited to time-bounded tasks and its frequency hopping scheme, to avoid interferers. Bluetooth allows very simple forms of networking. A network (in the Bluetooth jargon, *piconet*) is formed by a master and up to seven slaves, using asynchronous connection-less (ACL) mode. A point-to-multipoint full-duplex system can be achieved this way, with an effective data rate up to 721 Kbps. Packet retransmission is possible, at the expense of a lower effective data rate. Additionally, a synchronous connection-oriented (SCO) mode is available, which allows only point-to-point connection. A master can handle up to three SCO links of a constant data rate of 64 Kbps each, with guaranteed in-time data delivery. A more recent Bluetooth specification (v2.0) introduced enhanced data rate, reaching an effective data rate of 2.1 Mbps.

In [3] ACL was employed to create a point-to-point or point-to-multipoint system for compressed audio quality unidirectional streaming. The point-to-multipoint case only allowed two receivers to be implemented. The maximum measured distance was 10 m in the first case, and 3 m in the second one. Larger distances would require lower bitrates (i.e., lower audio quality) and would incur in packet loss. The Bluetooth broadcast mode was also used, but induced a large packet loss (over 2 %). Many other works dealing with Bluetooth and audio exist in literature, but are not of interest in this context or are of lesser quality. Bluetooth, although capable of some real-time music streaming is subject to limitations such as range, data rate, and latency. It does hardly scale, as piconets only allow up to 7 slaves, and more complex layering are required to introduce more slaves, increasing latency, networking complexity, and collisions. In conclusion, other networking technologies must be considered for NMP usage.

4.2 Proprietary Audio-Specific 2.4 GHz ISM Band Solutions

Commercial applications are available since years in the field of wireless headsets and loudspeakers, to deliver cinema audio or music contents. Some of these rely on commercial chipsets and standards (see e.g., Sonos[1] and Skaa[2]). These ad hoc solutions allow a product to differentiate from other ones in terms of reliability, bandwidth, and features and create a value proposition for vendors of the wireless equipment. IEEE 802.11 is also employed in this field as will be described later. The wireless technologies alternative to 802.11 may be appealing in terms of throughput, robustness to interferences, latency, and bandwidth. Skaa, for instance, is based on a proprietary protocol and system on a chip from Eleven Engineering Inc., which implements an adaptive frequency hopping algorithm (called Walking Frequency Diversity, patented), to dodge potential interferers in the 2.4 GHz ISM band, and

[1] http://www.sonos.com/.

[2] http://www.skaa.com/.

a set of constant latency values, down to a minimum of 10 ms, with a maximum latency offset between two receivers of 40 μs. The protocol, unfortunately, allows only for unidirectional signal transmission, although communication is bidirectional to collect RF link statistics and ask for packet retransmission.

Two silicon manufacturers that provide *wireless audio* solutions are Texas Instruments and Microchip, among others. TI provides the CC85x family of ICs, which operates at maximum 5 Mbps in the 2.4 GHz ISM band, avoiding interferers by adaptive frequency hopping and listen-before-talk, and providing forward error correction, buffering, retransmission to minimize errors. The audio links are CD quality, with wireless clock distribution (necessary to compensate for clock drifts) and maximum of four unidirectional audio channels are supported with a minimum latency of 10.6 ms (512 samples at 48 KHz). These characteristics make it unusable for direct usage NMP. Microchip branded the KleerNet technology that is targeted at unidirectional audio streaming to loudspeakers and wireless microphones. It operates in the 2.4, 5.2, and 5.8 GHz bands and guarantees an indoor range of up to 60 m. The marketed ICs host a large set of features, including audio processing and user input. The latency is declared to be less than 20 ms and clock sync may be achieved by means of digital interpolation. Again, this technology is not feasible for quick exploitation in wireless NMP.

4.3 Sub-1 GHz Large Bandwidth Solutions

Other custom solutions may employ the sub-1 GHz ISM bands for increased range. Analog wireless audio transmitter for stage use is traditionally in the VHF band. Analog or digital UHF equipment also exists and is widely used. The IEEE 802.11 task forces very recently proposed two standards in the sub-1 GHz range:

- IEEE 802.11af, ratified in February 2014, exploits the now unused VHF and UHF TV bands (53–790 MHz);
- IEEE 802.11ah, yet to be finalized, addresses low-power long-range applications, with limited bandwidth, ideal for Wireless Sensor Networks.

Of these, IEEE 802.11af could probably cover the needs for multimedia delivery, and extend the transmission range, with respect to other IEEE 802.11 physical layers due to the reduced carrier frequency. Channels are narrower, compared to IEEE 802.11a/b/g and n (6–8 MHz versus 20–40 MHz); thus, the theoretical throughput is reduced, but—depending on the employed modulation and other factors—yields throughputs ranging from 1.8 Mbps (6 MHz channel, binary shift-keying modulation, 6 μs guard interval) to 35.6 Mbps (8 MHz channel, 256 quadrature amplitude modulation, 2.25 μs guard interval) per channel. The protocol employs orthogonal frequency multiplexing and up to four channels can be exploited at once to increase the data rate. Multiple-input multiple-output operation allows to exploit four different spatial streams to increase data rate. The maximum theoretical throughput allowed by four channels and four spatial streams is as high as 568.9 Mbps for 8 MHz wide channels. The extended range and sustained throughput will probably make this standard a good candidate for wireless NMP as soon as computing devices support it. This protocol, however, supports a very wide spectrum, where broadcasts are still active, depending on the country. Cognitive radio techniques need be employed to avoid channels already in use by existing TV stations.

4.4 IEEE 802.11 ISM Band Protocols

As already outlined, a candidate for wireless NMP is the IEEE 802.11 family, nowadays regarded as the de facto standard for wireless communication in home, small office, enterprise, and institution scenarios. It provides, together with related standards, a means to deliver multimedia content and its data rates have been increasing with further published amendments. At the physical level, the 802.11a, b, g, n and ac protocols rely on the 2.4 and 5 GHz ISM bands. By employing different modulations, channels, and antenna diversity techniques, they obtain very different bandwidths and coverage ranges.

The 802.11a and 802.11 g are very mature and are of interest for their wide acceptance. They reach a maximum raw data rate of 54 Mbps, in the 5 and 2.4 GHz bands, respectively. The 802.11b first extended the 802.11a to the 2.4 GHz band and achieved a maximum raw data rate of 11 Mbps. Considering that the raw data rate only applies in the best case of transmission conditions and that error correction and overhead must be taken into account, the effective throughput is significantly lower. IEEE 802.11 is half-duplex; hence, for bidirectional audio transmission, the required bandwidth must be correctly estimated to fit the effective available throughput. In the case of multiple stations, the available bandwidth must be divided by the number of stations and extra overhead due to collisions and coordination must be considered.

The 802.11n amendment improves on the previous ones by achieving a data rate of maximum 600 Mbps by introducing MIMO antenna diversity over both 2.4 and 5 GHz bands. A last amendment recently implemented in devices and access points is 802.11ac, which further increases the data rate by increasing MIMO number, channels width, and modulation order. Typical maximum raw data rates obtained with this protocols are in the order of 1300 Mbps. Frequency range and data rates of all the aforementioned IEEE 802.11 standards are depicted in Fig. 4.1.

In a paper dating back to year 2007, IEEE 802.11b was employed in an experimental audio transceiver for guitars [4], where a 8-bit microcontroller was programmed to transmit audio packets acquired from a D/A. The results of the paper are barely in line with psychoacoustic requirements for NMP, as the link latency reported is of 15.8 ms. However, most of the latencies are due to implementation issues with the cheap integrated circuits and microcontroller employed, and more adequate components would not introduce the same transmission delay between integrated circuits. Furthermore, a very low data rate of 2 Mbps was used, the bare minimum to sustain 16-bit 44100 Hz audio transmission and related overhead. The broadcast capabili-

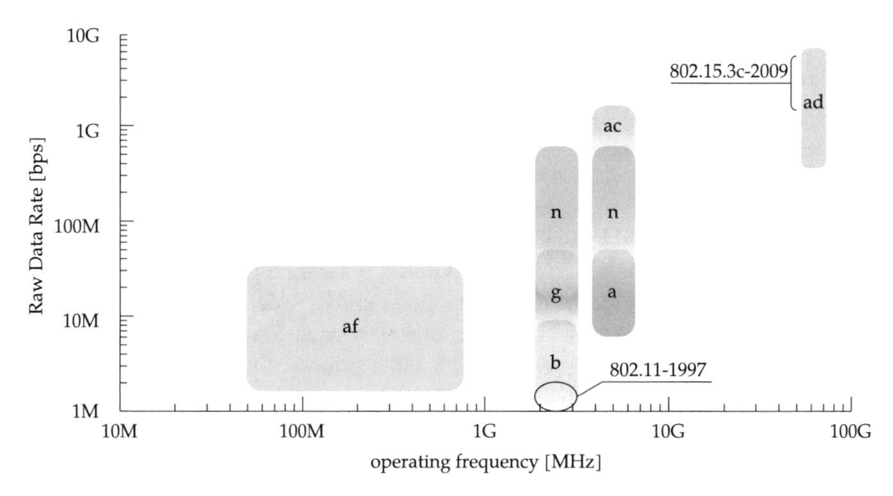

Fig. 4.1 Graphical comparison of IEEE 802 standards addressing operating frequency and raw data rates for operation in the 53–790 MHz range (802.11af), 2.4 and 5 GHz ISM bands (a, b, g, n, and ac), and 57–64 GHz EHF band (ad and 802.15.3c-2009). Please note that the data rates are dependent on the symbol rate, modulation scheme, and spatial/antenna diversity, and they are not representative of the actual throughput achievable. Commercial products may implement proprietary solutions to increase data rates, e.g., by employing multiple radio transceivers

ties of 802.11 were shown, allowing for point-to-multipoint networking and reduce overhead and collision, not requiring an acknowledgement from the receiver.

On the opposite, a very highly technical solution was employed in [5], where the potentialities of IEEE 802.11 were demonstrated providing a high-bandwidth streaming of audio with a very low latency (a unidirectional stream of eight audio channels at 192 KHz 32 bit with audio packets as small as around 50 samples was reported). IEEE 802.11 is half-duplex but allows bidirectional audio exchange; thus, the same bandwidth reported in that paper could be exploited for bidirectional audio connection.

Critical factors in using 802.11 protocols for NMP are as follows:

- the 2.4 and 5 GHz ISM bands are crowded by several communication technologies and electric appliances;
- medium access may sensibly delay audio transmission, especially with increasing size of the network;
- the hidden station problem may arise.

Regarding access to the medium, the DCF (Distributed Coordination Function) is a medium access protocol meant to be fair with respect to all stations, and it does not cope with time-bounded traffic. For this reason the PCF (Point Coordination Function) is also described by the standards, which allows a coordinator to schedule transmission during a contention-free period. The DCF and PCF are not mutually exclusive and can coexist in a superframe. A superframe includes a contention period and a contention-free period. More recently, an amendment to the standard IEEE 802.11e-2005 was proposed, introducing a new coordination function which enhances DCF and PCF, respectively, with enhanced distributed channel access (EDCA) and HCF controlled channel access (HCCA). EDCA introduces traffic priorities. The highest priorities are reserved to video and voice traffic, improving timeliness with multimedia content and videoconferencing traffic delivery. Other lower priority traffic sources are those often found in a home or office network, which are not constrained by real-time requirements. This distinction in traffic categories is hardly useful for NMP usage. A wireless network for NMP is very probably devoted to the unique scope of delivering audio and/or video packets for a performance. All audio or video packets have, thus, the same low latency requirements and possibly need to be assigned with the same priority. Only non-time-bounded traffic such as text messages between performers, network status, and information messages could be assigned a lower priority. If these are a consistent source of traffic and their transmission may temporarily steal some bandwidth or induce collisions, then EDCA is suitable.

Of higher interest for NMP is the HCCA. This protocol extends PCF by allowing contention-free periods to be called by the AP whenever it needs and by introducing traffic classes and priorities. A scheduler may be implemented and a maximum transmission time is imposed to each station. This flexibility allows very thorough planning of a network quality of service (QoS). Unfortunately, HCCA implementation is not mandatory and it appears to be very rarely present on access points or operating systems.

In [5], audio transmission is performed with 802.11n in the 5 GHz frequency band, at 108 Mbps data rate, employing antenna diversity. Each channel carries 32-bit audio at 192 KHz and is delivered using multicast. Audio is transmitted in the contention-free period using PCF. A contention period is available for additional data transmission. The contention-free period lasts 2.7 ms and is employed to transmit 12 audio blocks of approx. 50 samples each (approx 3 ms of audio). UDP multicast packets are sent and not require an ACK. Forward error correction is employed to recover from bit errors. In the prototype implementation two access points (built on desktop Linux machines) have been used to handle the high traffic. This work proves that IEEE 802.11 is worth considering for wireless audio streaming. It also shows that much low-level development must be done to make such a system work properly.

4.5 Millimeter Wave Band Multimedia Transmission Protocols

Looking at the extremely high frequency range (30–300 GHz), technologies are available which provide very large bandwidths, capable of multimedia transmission of audio and video. The IEEE 802.11ad defines a physical layer with channels in the 60 GHz ISM band. Propagation loss in this region of the spectrum is higher than other very common ISM bands. For instance, signal attenuation over the air at 60 GHz is of 68 dB per 1 m, while at 5 GHz it is of 46.3 dB. The transmission range at this frequency is clearly reduced with respect to the aforementioned standards. Advantages of this extremely high frequency are in the manufacturing of radiofrequency silicon components and the ceramic antennas, and the reduced range reduces collision, possibility of eavesdropping, or interference with the communication. Worldwide standardization of this frequency range has been along the way since 1994, when the US Federal Communications Commission first proposed to establish an unlicensed band at 59–64 GHz. Four channels are defined by ITU-R recommendations established together with industrial consortia fostering the development of this standard, namely the WirelessHD Consortium and the Wireless Gigabit Alliance. The WirelessHD consortium developed its own physical layer specifications. The specifications defined in 2010 allow rates up to 7.1 Gbps, further improved by a theoretical factor of 4 if employing 4 by 4 multiple-input multiple-output antenna configuration (MIMO). The protocol adopts a bidirectional low-rate physical layer (LRP) for link management and transmits unidirectional data over a medium- or high-rate physical layer (MRP, HRP). All the three defined physical layers (LRP, MRP, HRP) employ TDMA. Orthogonal frequency division multiplexing (OFDM) is always employed for this and other standards in the 60 GHz band.

The Wireless GigaBit Alliance, later merged into the WiFi alliance, follows the IEEE 802.11ad standard. This standard defines backward compatibility with IEEE 802.11 standards, and thus allows switching session between the legacy 2.4 and 5 GHz bands used in IEEE 802.11 and the 60 GHz band. Similar to WirelessHD, a low-data-rate channel is employed for bidirectional control and negotiation (MCS0), while high-data-rate modes are employed for data exchange (MCS1-31). The maximum data rate guaranteed by this standard is 6.75 Gbps.

Both the WirelessHD and the IEEE 802.11ad specifications only support *ad hoc* networking, which can be considered limiting for a flexible NMP system. Ad hoc networking allows only point-to-point communication.

Another set of IEEE standards address high-data-rate communication in the GHz range, in this case for personal area networks. IEEE 802.15.3 was standardized in 2003, and defined a physical layer with 11–55 Mbps data rates. A more recent amendments to this standard, IEEE 802.15.3c-2009, defined a new physical layer in the 60 GHz range [6]. Three main use cases are defined which in turn call for three different physical layer definitions. Of interest for NMP may be the high-speed interface mode (HSI), meant, e.g., for low-latency bidirectional high-speed data, as in conferencing. The audio/video physical layer, despite the name, is meant for asymmetrical multimedia content delivery, e.g., streaming from a signal source to a sink device, and thus is not appealing for NMP. Data rates achieved from HSI range 1.54–5.78 Gbps, and the expected range in practical usage may be of a few meters. Indeed, PAN protocols target very short range usage, and it is unlikely that RF device compliant with this standard will provide longer ranges and a higher output power with respect to the one suggested by the standard.

Figure 4.1 also includes some of the EHF solutions discussed above.

4.6 Wireless Transmission of Control Signals

Up to this point, wireless technologies for audio transmission have been reported. The outcome of this survey is that no prior work was found to employ wireless audio transmission for NMP. However, the wireless medium is employed nowadays for control data transmission by several manufacturers of MIDI (Musical Instruments Digital Interface) instruments through Bluetooth links or WiFi. Products like Pandamidi,[3] MidiJet,[4] PUC[5] are able to send MIDI data from one instrument to another, even bidirectionally at varying distance through the use of a couple of transceiver or from a MIDI controller to a smartphone or tablet device. These products generally guarantee very low latencies (e.g., 2.7 ms for the second product). Similarly, Hexler proposes since years a mobile device application, TouchOSC[6] to send OSC (Open Sound Control) data from the mobile device to another platform, e.g., a computer via WiFi. Often, in this case, a large part of the latency is due to the touch screen response time, as it was found during a study by the authors [7]. Also, being based on commercial access points, it is subject to common latency issue due to IEEE 802.11 medium access. All the aforementioned commercial applications do not implement networking.

The new interfaces for musical expression (NIME) community have been very prolific in wireless protocols for music interfaces. Bluetooth as well as ZigBee have been reported as a viable alternative to WiFi, especially given the very short range often required by music interfaces. Specifically, ZigBee has been outlined as a better alternative to Bluetooth for its improved connection stability and reduced latency [8]. The same authors state that IEEE 802.11 is the least prolific technology used in that field among the aforementioned three. Representative works employing wireless networking in the NIME community are [9–14].

Other uses of wireless networking for control data are sometimes found in the laptop orchestras' literature. Custom software written in Max/MSP, ChucK, Puredata, or supercollider is reported to support networking and sharing of information for the performance.

References

1. Reuter J (2014) Case study: building an out of the box Raspberry Pi modular synthesizer. In: Linux audio conference (LAC2014). Karlsruhe, Germany
2. Principi E, Colagiacomo V, Squartini S, Piazza F (2012) Low power high-performance computingon the beagleboard platform. In: Education and research conference (EDERC),(2012) 5th European DSP, vol 2012. IEEE pp 35–39

[3]http://pandamidi.com/.

[4]https://midijet.com/.

[5]http://www.mipuc.com/.

[6]http://hexler.net/software/touchosc.

3. Floros A, Tatlas N-A, Mourjopoulos J (2006) A high-quality digital audio delivery bluetooth platform. IEEE Trans Consum Electron 52(3):909–916
4. Jakubisin D, Davis M, Roberts C, Howitt I (2007) Real-time audio transceiver utilizing 802.11b wireless technology. In: SoutheastCon, 2007. Proceedings, IEEE, pp 692–697
5. Nikkilä S (2011) Introducing wireless organic digital audio: a multichannel streaming audio network based on IEEE 802.11 standards. In: AES 44th international conference, San Diego
6. Baykas T, Sum C-S, Lan Z, Wang J, Rahman M, Harada H, Kato S (2011) IEEE 802.15.3c: the first ieee wireless standard for data rates over 1 Gb/s. IEEE Commun Mag 49(7):114–121
7. Gabrielli L, Squartini S (2012) Ibrida: a new DWT-domain sound hybridization tool. In: 45th AES international conference, Audio Engineering Society
8. Mitchell T, Madgwick S, Rankine S, Hilton G, Freed A, Nix A (2014) Making the most of wi-fi: optimisations for robust wireless live music performance
9. Fléty E (2005) The wise box: a multi-performer wireless sensor interface using wifi and osc. In: Proceedings of the 2005 conference on new interfaces for musical expression, pp 266–267
10. Fléty E, Maestracci C (2011) Latency improvement in sensor wireless transmission using IEEE 802.15.4. In: Proceedings of the international conference on new interfaces for musical expression, pp 409–412
11. Overholt D (2012) Musical interaction design with the cui32stem: wireless options and the grove system for prototyping new interfaces. In: Proceedings of international conference on new interfaces for musical expression, Ann Arbor
12. Jenkins L, Page W, Trail S, Tzanetakis G, Driessen P (2013) An easily removable, wireless optical sensing system (eross) for the trumpet. In: Proceedings of the international conference on new interfaces for musical expression (NIME 2013), Seoul, Korea, pp 27–30
13. Madgwick S, Mitchell TJ (2013) x-osc: a versatile wireless i/o device for creative/music applications. In: Sound and music computing conference (SMC2013)
14. Torresen J, Hafting Y, Nymoen K (2013) A new wi-fi based platform for wireless sensor data collection. In: Proceedings of the international conference on new interfaces for musical expression

Chapter 5
Wireless Networked Music Performance

Abstract Wireless NMP has very few examples in the literature, if none (depending on the definition the reader adopts for NMP). This chapter reports advancements and developments in wireless NMP. The challenges posed by wireless NMP and the opportunities it offers are different from those seen in wired remote NMP. For this reason, a specific approach and framework has been developed by the authors which is reported in this chapter and compared to other meaningful approaches and technical achievements from other authors in wireless and wired NMP. Most of the contributions are by the authors and colleagues. The rationale and goals of the authors' project, named WeMUST, are described and its technical achievements later reported. The project also targets portability and ease of use in wireless NMP. Embedded platforms are, thus, employed which are power-autonomous and provide some DSP capabilities. They adopt connection automation tools based on custom service discovery mechanisms based on existing networking technologies. The software used and related parameters are described and motivated. Finally, issues related to outdoor use are reported and technical choices to overcome these are described.

Keywords WeMUST · Wireless networked music performance · Embedded computing · Clock synchronization · Automatic connection · Service discovery · Linux

5.1 The Wireless Music Studio Project

The framework developed by the authors has been called wireless music studio, in short WeMUST, and has been conceived to explore the possibilities offered by wireless networking in the music performance and studio context with a special focus on technical challenges and their possible solution with widespread hardware and protocols. Of specific interest is the use of embedded platforms in substitution to the laptop as a musical and interaction instrument.

The key concept underneath the investigations within the WeMUST project is the use of wireless transmission for indoor and short to medium range outdoor locations and IP networking to enable audio and control signal transmission in music recording and performance. Neither the transmission of wireless audio nor the use of IP

L. Gabrielli and S. Squartini, *Wireless Networked Music Performance*,
SpringerBriefs in Electrical and Computer Engineering,
DOI 10.1007/978-981-10-0335-6_5

for audio networking are totally novel: point-to-point radio technologies have been available commercially since a few decades for large stage usage and networking is currently employed with broadcasting studios as reported in Appendix A. However, the adoption of general purpose communication technologies (in this case IEEE 802.11 and IP networking) in challenging scenarios such as those under investigation within the project have been scarcely addressed by academic research, and is still to be considered by the industry. The reason for the latter may partly be due to a negative bias among musicians and technicians toward digital wireless technologies reliability, reducing the commercial potential of new products featuring such technologies. A common-ground experience regarding wireless networking is most probably that of personal computer and mobile devices relying on 802.11 protocols. While this set of protocols is mature and globally adopted, users may still experience troubles due to its complex software architecture, to signal coverage, interference, channel crowding, etc.

5.1.1 Application Scenarios

By taking advantage of wireless communication and portable platforms, several application scenarios are possible:

- Networked laptops for ensemble performance
- Networked embedded platforms or mobile devices for signal acquisition, processing, and transmission
- Acquisition devices to capture audio and send to a DAW or a public address system
- Controllers to send data to a synthesis engine or effects processor
- Body sensing for musical or dance performance transmitting information to a processing unit
- Wireless loudspeakers

Home or small studio music production systems can be envisioned. In such a scenario musicians can setup their instruments in the live room, turn their transceivers on and quickly connect to WeMUST-enabled amplifiers, loudspeakers or digital mixing console. The signal can be routed at the same time to a digital mixer for recording in the monitoring room. In a studio all the instruments including MIDI controllers can be routed to a PC acting as DAW (Digital Audio Workstation). The computational workload can be distributed among several devices in a network, relieving the DAW running on a PC from part of the processing, following recent experiments with LAN [1, 2]. For instruments such as guitars which require effects to be applied, either the clean signal from the instrument either the output from the amplifier can be streamed to the recording mixer. In the former case applying more specific effects is left to postproduction while the musician can perform with a sound he feels comfortable playing with. If a technician is not required the musicians can control the digital mixer in the monitoring room together with the multitrack recording gear with a tablet device directly from the live room.

In live contexts, the time consuming task of deploying cables from the front and stage mixers to the stage, instaling microphones, and sound checking can be avoided and made more reliable with the use of a wireless technology that allows to quickly and flexibly connect the sound sources to the mixers. No check need to be performed after the devices are connected and each instrument can diagnose the network reliability and bandwidth with ease by automatic software checks.

Common USB musical keyboards and controllers can thus convey the MIDI or OSC data to the PC wirelessly. Regular expander racks and keyboards can also be fitted with a WeMUST transceiver to send audio data directly to the PC. If they are fitted with hardware controls or a touch surface they can control the mixing parameters. The same wireless architecture can be also employed for control of live lighting, smoke, and fog.

Devices that transmit and receive audio can also manipulate audio. The envisioned hardware platform must be, thus, also capable of performing DSP.

Avoiding cables may also reduce the risk of ground loops that introduce hiss and hum into the audio equipment and avoid risk of accidental tripping over cables.

Drawbacks must be highlighted as well. In that regard, informal conversations occurred in several professional and academic contexts with the authors. The *linux-audio* community mailing list provided several critiques in the last two years, including latency, cost, and reliability of the solution in terms of dropouts. Reliability as with any kind of wireless communications is related to time-varying fading and multipath which decrease the signal quality, especially with moving people or objects. Furthermore, wireless technologies are nowadays of public concern for health-related issues, hence attention must be paid to reduce as much as possible the amount of radiated energy from devices in a given space. Directional antennas may help in reducing the radiated energy.

5.1.2 System Topology

As a whole, WeMUST is currently a set of open-source software tools designed to fulfill the requirements specified in the introduction of this chapter. The tested and supported HW is a development platform called Beagleboard xM (BBxM), based on an ARM Cortex-A8 core, and any x86-powered personal computer running Debian or similar Linux distributions. Additional HW that may be required for specific purposes is detailed in Sects. 5.7 and 6.1.1. Other audio equipment such as high-quality sound cards can possibly be employed but will not be discussed here. A WeMUST system can be, thus, built with rather inexpensive hardware.

At the audio level, the networking paradigm is that of a peer-to-peer network, where every node can act as a transmitter and receiver, with no hierarchical pattern. This is required to allow for the maximum flexibility of the network topology that only requires two nodes of any kind to be active in the network. For this reason,

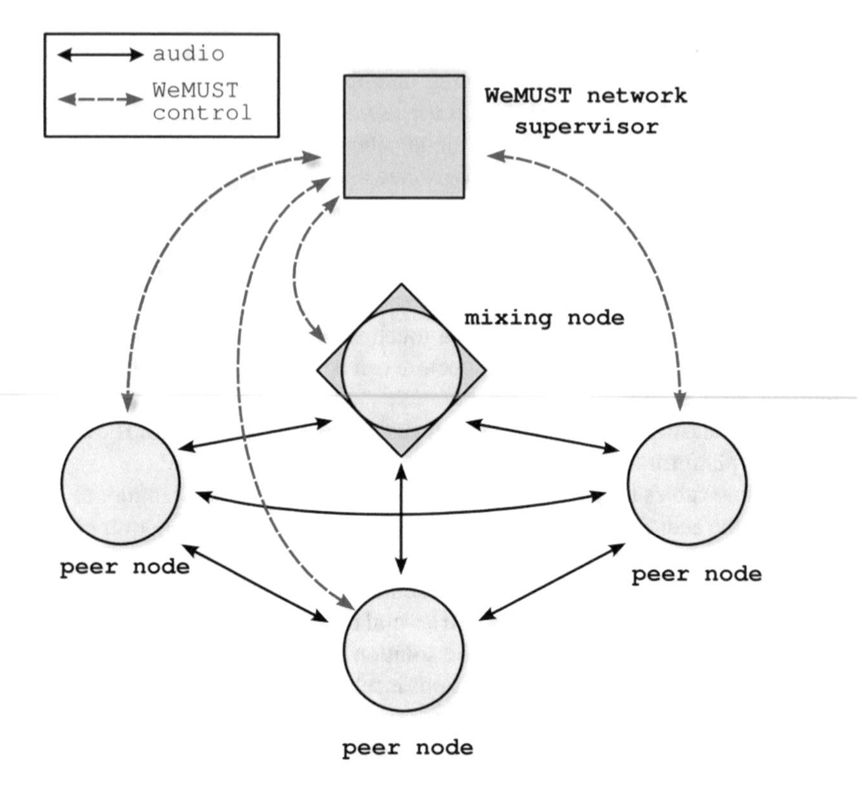

Fig. 5.1 Schematic view of a WeMUST network comprising three nodes and a mixing node mesh connected. The number of nodes and their connection can be arbitrary. A supervisor can scan the network for devices, control the devices and their connections, receive debug messages

any master–slave scheme, as this enforced with other software tools[1] is disregarded. Nodes of this network can independently enact any of the following: acquire and/or emit acoustic signals, transmit and/or receive digital audio or control signals, record and store signals. Nodes receiving signals from multiple nodes and/or recording them are called *mixing* nodes (in short MPCs in the case of personal computers). Nodes can also monitor the network for debug purposes and issue commands to other nodes for connection, shutdown, activation, etc. These nodes are called *supervisors* and are not necessary if the other nodes are programmed to act independently or are controlled by the user from a user interface (e.g., buttons). A simple diagram exemplifying a WeMUST network with three nodes, one mixing node, and a supervisor is depicted in Fig. 5.1.

An example setup is shown in Fig. 5.2, where an MPC can transmit control data (e.g., coming from a DAW or an algorithm), a click track for musicians synchronization, and receives audio data for recording. The nodes can capture or synthesize audio, or only be employed for remote MIDI control.

[1] Such as *netjack*, the network driver built in JACK.

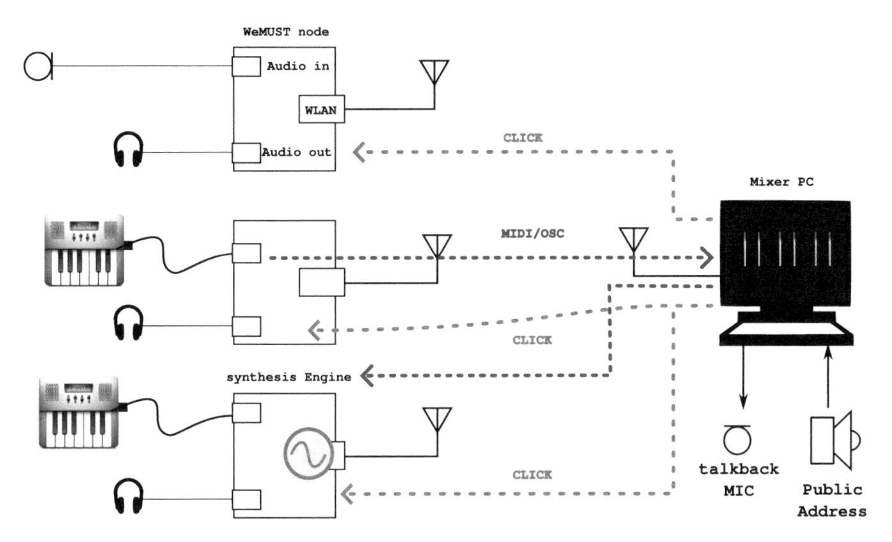

Fig. 5.2 Example of a WeMUST setup, involving audio, and control data transmission between embedded devices and an MPC

Each node must be connected to the network through a suitable wireless or wired interface, currently a WiFi transceiver or an Ethernet controller. Commercial wired routers and/or wireless Access Points are required. To establish a connection with other nodes connection parameters must be configured by the user. A script is provided to help the user configure the OS during installation. Implementing the SABy protocol [3], devised for use in WeMUST similar contexts, facilitates connection. SABy is based on a simple multicast and unicast messaging system (conceptually similar to UPnP) and DNS.

The nodes can run WeMUST OS, a specific Debian image customized for the goal at hand. WeMUST OS already contains all the packages required for the applications targeted by the WeMUST project. As such, all the software is based on GNU/Linux and a collection of standard tools required for audio and networking, supplied in the WeMUST OS image. Most of the software is written in Bash and Python scripting languages for ease of use, distribution, and update. Jacktrip, a cross-platform NMP client by the SoundWire research group at CCRMA Stanford [4] is supplied for the audio data transmission.

5.2 Automating Connections and Audio Networking

In the path to a seamless audio connectivity framework some effort is required in order to provide metadata regarding the devices and session control for the audio and video streams. In the field of wired Audio over IP protocols these issues are well addressed and allow for a set of functionalities specifically meant for routing audio from and to devices, controlling parameters of audio devices in the network, synchronizing clocks, and other timing information among devices in the audio network and ensuring flawless audio transmission between ends. A NMP requires similar features. The academic literature provides some tentative implementations, but no real convergence has been found at the moment on the requirements and the way to implement these.

5.2.1 Automatic Network Configuration

In computing, configuration is a first step that is required to take access to the network resources. This often requires technical skill and prior knowledge on the network topology and the services it provides. In IP networking all devices wishing to connect to a certain domain need to have an IP address assigned and have further information of the network (e.g., netmask, gateway address, etc.), which is necessary to establish connections to other machines. The configuration may take time if performed manually and requires a certain knowledge of IP networking. Furthermore, manual intervention with modern computing devices is more difficult, considered the relatively small-sized human–computer interfaces of portable devices or the lack of a proper input interface. Recent home automation and Internet of things devices, can operate through a LAN or a WLAN and can be installed permanently in a home or a building without further intervention. Such devices (thermostats, fire alarms, sensors, wireless music streaming devices, etc.) may have close to none input interface. For this reason, standards and technologies have been proposed and are nowadays implemented in a large number of devices. It is the case, e.g., of the Wi-Fi Protected Setup (WPS), a functionality implemented in most 802.11-enabled devices and access Points. This allows to connect the two by the press of a button the two devices have and thus removes the need for any alphanumerical interfaces at the devices wishing to join the network. Once the device gains access to a 802.11 network at the OSI layers 1 and 2, to further simplify configuration of layers 3 and 4 other technologies are provided, such as Zeroconf. This is an umbrella term encompassing several protocols for automatic device configuration and discovery, currently employed for general purpose computing. It is based on dynamic host configuration protocol (DHCP) and domain name system (DNS) for automatic IP address assignment and resolution of hostnames, i.e., translation of hostnames into IP addresses and viceversa. Multicast addressing is a key ingredient in the automation of networking, as it relieves a device from keeping track of the exact network topology and the connected devices. For the purpose, one of the most widespread protocol is RFC 6762 [5], or multicast DNS (mDNS), which allows DNS operation without a specific DNS server.

5.2.2 DNS-SD and SSDP

mDNS and DNS do not provide any description of device or the services it provides. Other protocols built on top of this are needed for the purpose. One protocol often employed on top of mDNS and DNS is DNS-SD (RFC 6763) [6], which provides device information over DNS messages:

- DNS SRV (RFC 2782) for service instance description
- DNS TXT RDATA (RFC 1035) for the same scope
- DNS PTR (RFC 1035) for querying for services on the network

The Audio over IP protocol RAVENNA employs Zeroconf and DNS for configuration, and DNS-SD for device discovery. Dante employed DSN-SD in previous implementations, while more recently it changed to a proprietary system.

Another very frequently used protocol for home multimedia networks is the simple service discovery protocol (SSDP), used by the UPnP Forum,[2] on which DLNA-compliant products are based. The UPnP set of protocols is implemented for many multimedia devices, such as smart TVs, smartphones, PCs, NAS (Network Attached Storage) devices, etc.[3] Both open and proprietary implementations exist, ported to many platforms. The UPnP standard is very complex and targeted to a large number of applications: from video streaming to home audio distribution, from network printer access to multimedia content browsing. Audio streaming in UPnP is handled as a reliable compressed stream, with large buffering and thus, high latency. As such, it is not targeted to music production or performance use. The SSDP layer, however is, of interest for several reasons: based on IP standards, it provides a server-less discovery mechanism. SSDP is a HTTPU text-based protocol encapsulated in UDP datagrams which are sent to a multicast address for discovery or to unicast addresses for information exchange. The SSDP and upper UPnP layers guarantee devices not only to discover other UPnP devices but also to describe their set of features and resources and have access to them.

Similarly to DSN-SD, SSDP includes three types of messages: advertise (uses the `NOTIFY * HTTP/1.1` header), search (uses the `M-SEARCH * HTTP/1.1` header), and search response (uses the `HTTP/1.1 200 OK` header). The advertise message is issued by a root device for different purposes: announce a new root device connection or reaffirm its presence (Notification Sub Type: `ssdp:alive`), several NOTIFY messages must be sent for each resource or service), change, or update information about the set of features and services available (Notification Sub Type: `ssdp:update`), announce the removal of a resource or service from the network (Notification Sub Type: `ssdp:byebye`).

The general connection scheme implies announces for a root device when it first connects to the network, one for each of its embedded devices or services. Control points can issue a search to the entire network or a unicast address in order to receive

---

[2]www.upnp.org.

[3]UPnP Forum "UPnP Specifications v.1.1", available online at: http://upnp.org/specs/arch/UPnP-arch-DeviceArchitecture-v1.1.pdf.

further information on its capabilities. This sort of information is described in XML format. The device description part of the UPnP is outside the scope of the device discovery and will not described further.

5.2.3 Open Sound Control

Open Sound Control (OSC) is a content format for musical purposes and such [7]. It prescribes a message formatting useful for description, content, and timing information exchange between software synthesizers, mixing consoles, etc. The specification is open and is often seen as a modern replacement for the musical instrument digital interface (MIDI) protocol. This, although, is not correct, as OSC has a much wider extent, little standardization, and provides no real communication protocol specifications. For this reason MIDI and OSC coexist. OSC has a URL-style symbolic naming scheme, in order to access services or components of a device in a tree structured. Taking pace from this feature, Eales and Foss proposed and demonstrated a device discovery system [8] based on OSC. Other extensions to OSC are proposed in [9], with, e.g., an interoperability feature in the URL scheme guaranteed by dividing the tree in three branches: public, vendor, and private. Private may be any message that does not require interoperability and is meant for personal use. Any OSC message conforming to current OSC specifications would fall in this category. The new public and vendor categories would partition the OSC URL root into `/public` and `/vendor`. The public address scheme would include common features required by most users and decided by a committee, while vendor would contain proprietary methods and resources needed by instruments and interfaces vendors, institutions, and software developers. Another interesting addition is the inclusion of MIDI commands into a `/public/midi` branch. In [10], a networked ensemble software is proposed, Espgrid, which is partially based on OSC and employs broadcasting a *beacon* message when a new instance of the software is created. The beacon does not only serve for announcement but also for estimating latency between two devices in the network.

Although the aforementioned works propose service discovery mechanisms, the OSC 1.1 specification, previously published in 2009 [7], proposed a very basic discovery mechanism, based on DNS-SD.

5.2.4 Open Control Architecture

As of December 2014, the open control architecture (OCA) alliance, supported by several audio equipment companies, released specifications for a control, and monitoring protocol of audio media networks. This protocol provides device discovery, media streaming management, control and monitoring of devices and software or firmware udpate of the devices. It provides security features, network and link control, device, and network diagnostics. The OCA is object-oriented and provides a large number of methods to control parameters including gains, filters, signal generators, dynamics processors, etc. It provides specifications for the upper three layers of the OSI model, and can be used in conjunction with existing media transport protocols. The specifications are expected to be ratified as a new AES standard due year 2015 and the current draft is available online.[4]

5.2.5 Simple Autonomous Buddying

Although the aforementioned allow for device discovery to various degrees, in the development of WeMUST a system was built from scratch and adapted to the use cases until a satisfying solution was found. While protocols such as SSDP are quite flexible it provided an option even too complex for audio networking. The authors proposed a simpler solution, partly inspired by SSDP, called simple autonomous buddying (SABy), implemented in the Python scripting language and as a Puredata C external. The Puredata external implementing SABy has been called `[netfind]` and works together with two modified versions of `[netsend~]` and `[netreceive~]` originally designed by Olaf Matthes.[5] The audio streaming is handled by the latter two externals, while the device discovery is handled by `[netfind]`, which configures the two streaming externals for proper connection to and from a new peer, when found. In the Python implementation SABy-related classes and functions are available in the wemust.py library, however, the code is embedded into a Python GUI (later described in Sect. 5.2.5.2) for device control and connection. The Puredata implementation is, thus, less goal-specific and shall be taken here as example to detail the SABy specifications.

SABy is similar to SSDP, but stripped down for the application at hand. A multicast group (a multicast IPv4 address, i.e., 239.255.255.251:1991) is used, similarly to SSDP for announcement of new devices. `[netfind]` can instantiate four different child threads:

1. `netfind_listen`: listens to announces sent to the multicast group. When a valid announce is received instantiates a new `netfind_flowctrl` thread.
2. `netfind_announce`: announces the existence of the device to the multicast group at the instantiation of `[netfind]` and repeats the announce at a certain interval `ANNOUNCE_INTERVAL` of several seconds.
3. `netfind_flowctrl_listen`: waits for incoming TCP connections at the port `FLOWCTRL_TCP_PORT`. It can accept multiple connections. For each connection negotiates the audio parameters, ports for audio streaming, etc. If the peer `netfind_flowctrl` thread accepts the connection, they keep the TCP connection alive and they check on each other's state by requiring a status message and reply with a *ACK* message.

[4]http://ocaalliance.com/.

[5]http://www.nullmedium.de/dev/netsend~/.

4. `netfind_flowctrl`: is started when `netfind_listen` finds a suitable peer to connect to. It tries to connect to that peer on its `FLOWCTRL_TCP_PORT` port and initiates negotiation with it. The negotiation includes a check for the match of the audio parameters (at the moment the sampling rate only) and then tries to agree with the peer for port numbers for incoming and outgoing audio streaming. If successful, it sends connection parameters to the `[netsend~]` and `[netreceive~]` externals connected to `[netfind]` outlets.

In Fig. 5.3 the interaction of two devices is illustrated, with L and F being Peer1 `netfind_listen` and `netfind_flowctrl` child threads, while A and FL are Peer2 `netfind_announce` and `netfind_flowctrl_listen` child threads. Each peer have the identical set of threads, however for illustration purpose only the ones involved in a handshake are shown. For instance, Peer1 has its own `netfind_announce` and `netfind_listen` threads, but since the two peers appear at different instants, the first one to spot the other's presence is Peer1, which occurs to receive Peer2 multicast announce and starts the negotiation between the F and FL threads. On the other hand, the Peer2 `netfind_listen` thread will spot the presence of Peer1 only after their connection, so that Peer2 `netfind_listen` will not start a new child thread for connection with Peer1. The `ANNOUNCE_INTERVAL` time must be high enough to statistically avoid two peers to simultaneously start a connection negotiation but must be low enough for the user experience to be responsive enough. For a simultaneous connection negotiation to happen from both sides the two devices must send their announcement messages in a time interval shorter than the connection negotiation duration (time from the creation of thread F1 to success). This time interval generally takes from 5 to 6 ms. This condition thus can only happen when two devices are both started in this short-time interval. The simultaneous connection negotiation can thus avoided by using on each peer a register accessible from all its child threads to track peers with whom connection is in negotiation. On the other hand, to improve the connection responsiveness a search can be issued by `[netfind]` at start instead of waiting for announcements.

The announce message is a text packet of the form shown in the following.

```
APP:<application name>
SR:<samplerate|period size>
NID:<ID>
TAG:<optional>
```

Listing 5.1 Announce message format

The first three fields are mandatory. The `APP:` field stores the application (e.g., puredata) that implements the protocol. The `NID` field can be a user-provided name or a unique 10-character alphanumeric ID assigned by the system at the instantiation of `[netfind]`. The `SR:` field stores audio paramenters, while the `TAG:` field is optional and can store keywords describing the device that can be searched for. Multiple connections can be handled and initiated this way very easily. It is up to the user to prepare the Pure Data patch in order to handle audio from the peer platforms.

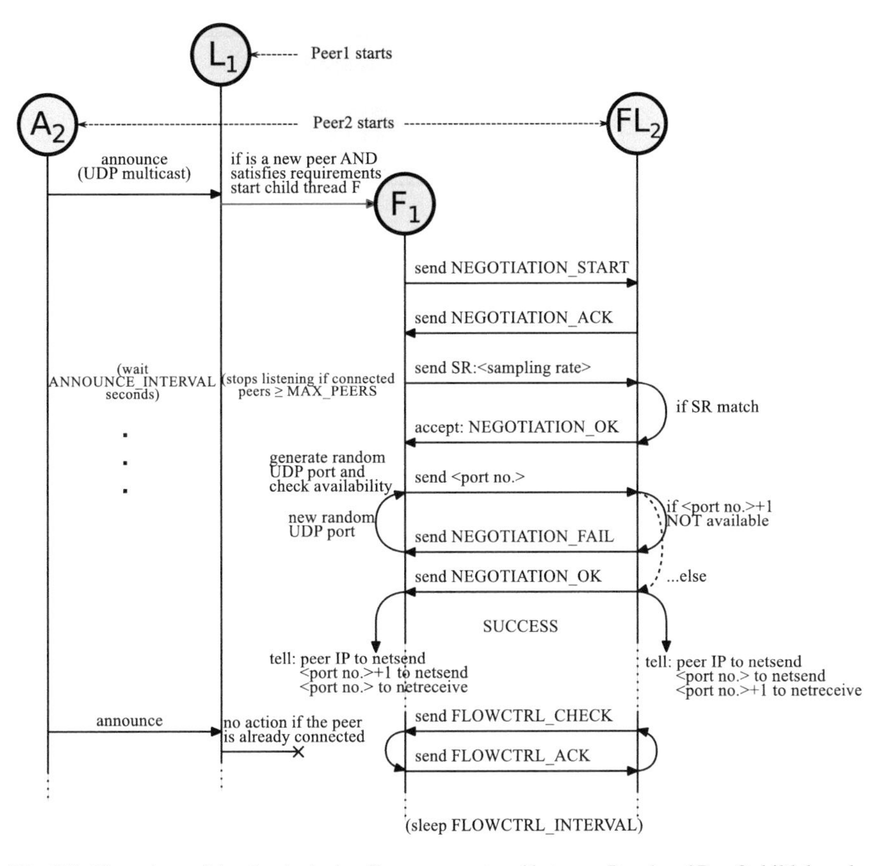

Fig. 5.3 Flow chart of the simple device discovery protocol between Peer1 and Peer2 child threads. L stands for Peer1 `netfind_listen` thread, A and FL stand for Peer2 `netfind_announce` and `netfind_flowctrl_listen` threads, F stands for Peer1 `netfind_flowctrl` thread

An example patch is provided together with the externals source code[6] that is able to control several `[netsend~]` and `[netreceive~]` externals, by the way of multiplexing netfind messages. In Fig. 5.4 a 2-way communication implementation based on a single Pure Data patch running on two different machines is illustrated for clarity.

Selection of peers to connect with can be also done based on a list of capabilities. When a `search <tag>` message is issued to `[netfind]`, it performs a query to the multicast group for devices that are tagged accordingly. From that moment on it will also discard announce messages that do not contain the specified tag. Tags are added to a `[netfind]` external by issuing a `addtags <tag list>` message. The tagging mechanism allows for quick selection and connection of available devices. For instance, a guitar stompbox based on Pure Data patches

[6]http://a3lab.dii.univpm.it/research/wemust.

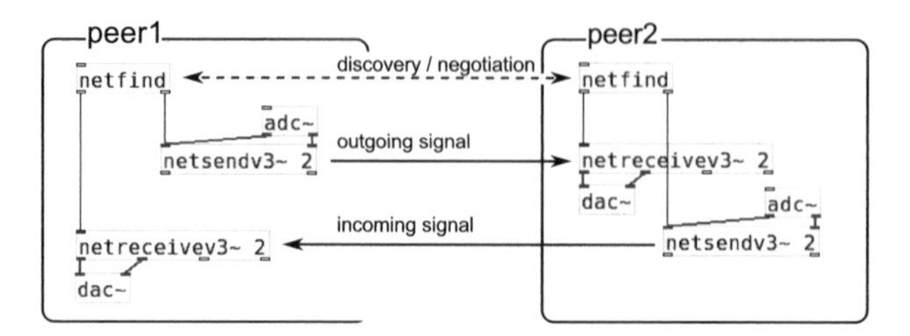

Fig. 5.4 A 2-way communication implemented in Pure Data using [netfind]

and [netfind] can be configured to only connect the input [netreceive~] external to devices tagged as "guitar," and the output [netsend~] external to a device tagged as "mixer" without need of human intervention.

5.2.5.1 Comparison with Other Protocols

The SABy mechanism described hereby inherits some concepts from SSDP. It is however different in some respects. Two design principles have been taken into account: flexibility mediated by small computational footprint. A complete UPnP software stack could be employed and adapted to the application at hand and development could be based on existing open C/C++ libraries. However, this would create a high dependency on rather large libraries, decreasing portability, which results unnecessary and would create large compiled binaries for such a simple task. This lightweight protocol includes all primary information in the announce message. Description of the devices is based on a text tags and a simple search/query mechanism, instead of the much more complex UPnP Description layer, which in turn requires an additional XML parser library. Overall responsiveness is very high, as mentioned above and the source code is quite compact and self-contained.

The parameters employed by SABy turned out to be almost the same as those employed by Dante, although at the time SABy was developed, Dante had not been investigated yet and the authors were not influenced by its specifications.

5.2.5.2 Prototype Implementation of Connectivity Software

The way discovered data is represented and proposed to the user is equally important as the way it is found on the network. Two GUIs were developed by the authors in a iterative fashion to investigate on this aspect. The GUI were designed in Python and GTK+[7] to allow for a flexible design. The GUIs served as a counterpart to the Python scripts running on the embedded audio devices, which had no input interface

[7] http://www.gtk.org/.

(besides a push button). The scripts developed for the embedded BBxM platforms were:

- jackaudio
- wemust-daemon
- read-usrbutton

At boot time the latter two are started by an `init` script. This script is started when networking services, the frequency scaling service, and all the other required functionalities have been started.

On a supervisor machine other Python tools are required: wemust-netdbg and wemust-connect (both spawn a different GUI). Both the BBxM and the PC scripts are based on a Python library of function, wemust.py.

Figure 5.5 reports a simplified diagram that clarifies interactions and functionalities of the scripts running on the BBxM. When all the required system services are loaded, an `init` script is started that loads the wemust-daemon and the read-usrbutton daemon. The latter controls the user button available on the BBxM, exposed as a `sysfs` item. Two modes are available: the button can act as a reset button (each time it is pressed jackaudio is stopped and restarted), or as a toggle to start and stop jackaudio. Wemust-daemon takes care of two parallel threads that will run until the system stops. The first, `getTempThread` periodically reads the CPU core temperature and sends it to the supervisor PC. The second, `ctrlThread` waits for commands from the supervisor PC, and executes them, sending back to the supervisor any command output. This is used to control the BBxM remotely or request for status updates (e.g., running processes, or kernel logs).

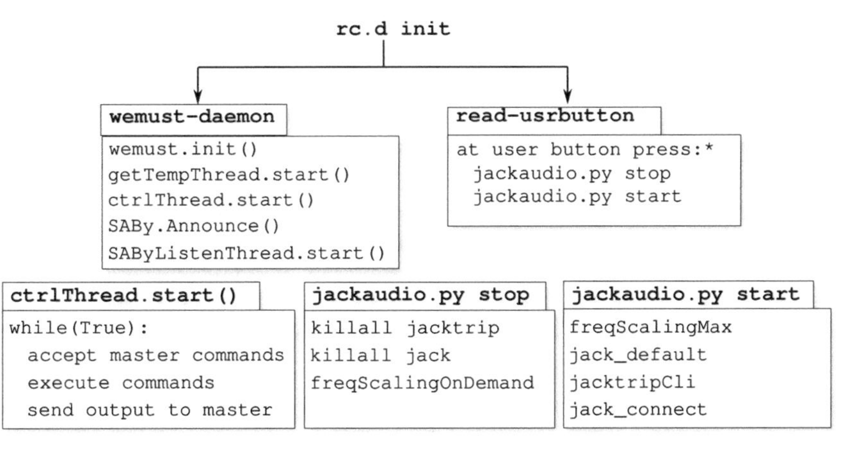

Fig. 5.5 A diagram showing functional sections of the scripts running on WeMUST OS after boot

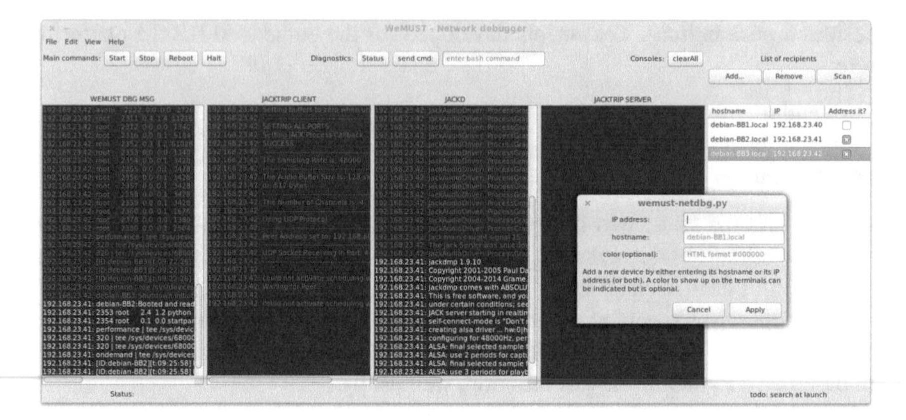

Fig. 5.6 wemust-netdbg GUI

When either the user button is pressed, or the supervisor PC issues a "Start" command, jackaudio is started, which will set the frequency scaling to the maximum core frequency, start the JACK server, launch Jacktrip and connect it to the input and output. All the options to JACK, Jacktrip and the connections can be set by editing the fields at the top of the script or passing command line options (as done by wemust-connect).

On the supervisor PC, the scripts present a functional GUI, as reported in Fig. 5.6. The problem of presenting the supervisor PC user with log messages coming from multiple IPs and multiple applications requires a markup strategy. Multiple consoles could be created, one for each peer. However, in this case, the higher the number of peers, the higher the number of consoles, with the risk of crowding the PC screen. A functional distinction has been made instead, allocating one console per function, i.e.,: general logging information, messages from JACK, and Jacktrip. Stacking these consoles horizontally allows to follow easily the time flow from top to bottom. Different colors are assigned to different peers, in order to increase readability.

The current GUI for wemust-connect, depicted in Fig. 5.7, takes inspiration from the connection utilities in QJackCtl,[8] a graphic frontend for JACK. The graphic metaphores are the same, however, in this case, peers have ports that are connected with other peers. The connection takes place by sending appropriate console commands to the peers through a SSH connection. By implementing a session control protocol, the Start, Stop, and other commands could be issued without requiring SSH access to a Unix/Linux machine and making the system more portable.

[8]http://qjackctl.sourceforge.net/.

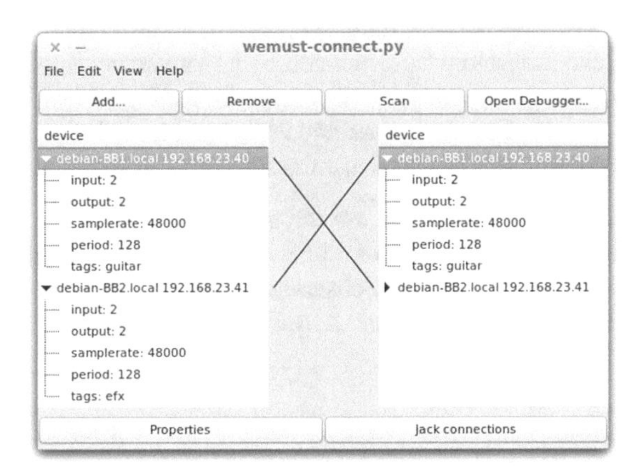

Fig. 5.7 wemust-connect GUI

5.3 Clock Synchronization

In Sect. 3.7 the clock synchronization issue was introduced, together with an adequate formalism. The communication technology literature provides several solutions depending on the use case constraints. As an example, the CS-MNS algorithm is provided in Sect. 3.7. Different approaches have been tested as part of the WeMUST project. Absolute time approaches have been discarded as motivated in Sect. 5.3.1 and two different relative time approaches have been tested with success during WeMUST experiments.

5.3.1 Absolute Time Approaches

Several approaches employed in wired Audio over IP, are based on aboslute time approaches. They make use of , e.g., the Precision Time Protocol (PTP) and similar ones. Dante, e.g., employs PTP (IEEE 1588-2002) to keep devices synchronized as close as 1 μs to a master clock. RAVENNA, similarly employs PTPv2 (IEEE 1588-2008) [11]. The PTP is based on a set of underlying hypotheses to estimate the offset between two clocks, a master and a slave. The master clock is selected among the clocks available in the network according to best master clock selection algorithm. Each slave needs first to perform a *syntonization*, i.e., what is called a relative clock synchronization. In order to do so, the master periodically transmits messages to the slaves at the rate of the order of the Hz. The slave adjusts its clock pace according to the periodicity introduced by the master by means of a control loop. Absolute synchronization is achieved by a round-trip time estimate to estimate the offset ρ (as defined in Sect. 3.7) between the two clocks. The mechanism require the exchange of three messages in the following order:

- The master sends timestamp T_M^1. The slave receives it and associates reception to the timestamp T_S^1 generated by its local clock.
- The slave sends back a message indicating its own current timestamp T_S^2 which is received by the master and associated to its local timestamp T_M^2.
- The master finally transmits T_M^2 to the slave.

At the end of the process the slave knows T_M^1, T_S^1, T_S^2, and T_M^2. By imposing the hypothesis of a symmetric network (i.e., the same delay n applies to messages in both directions) and by observing that $T_S^1 = T_M^1 + \rho_{SM} + n$ and $T_S^2 = T_M^2 - \rho_{SM} + n$, the offset can be estimated by evaluating

$$\hat{\rho_{SM}} = \frac{T_S^1 - T_M^1 + T_M^2 - T_S^2}{2}. \tag{5.1}$$

An accurate relative synchronization is required as hypothesis, and is achieved by means of syntonization, to impose that the offset does not change sensibly in this short time frame.

After absolute synchronization is achieved, and the slave system clock has been updated, the slave needs to update the audio card clock accordingly. In some cases the reference clock can generate or align the audio clock by means of dedicated hardware circuitry, similarly to what is done by a Word clock in professional audio equipment. However, as Weibel states in [11]:

> In the absence of hardware timestamping support, the principle can also be implemented in software, but the timing uncertainties typically inherent in software execution will require sophisticated filtering to achieve acceptable accuracy.

This is the case with PC-based nodes. Accuracy of the slave clock synchronization strongly depends on the aforementioned assumptions and the precision of the timestamp. Specifically, in the encapsulation of PTP messages, the OSI layers are descended from the application layer to the physical layer. The PTP payload is encapsulated in several containers (UDP packet, in an Ethernet packet, etc.) and is processed with some delay at both the transmitting and receiving ends. The instant when the timestamp should be taken and inserted into the packet as PTP payload should be as close as possible to the physical layer, to avoid the unknown , and potentially random time it takes for the packet to proceed through the OSI layers and be transmitted. For this reason, the network interface hardware should support PTP natively. With personal computers, the control over this hardware feature is not straightforward. Furthermore, since audio clock synchronization would anyway require some filtering, the authors suggest that DSP-only approaches for audio networking with personal computers is preferable.

In NMP, specifically, only relative synchronization is required. The time functions of two remote ends may have a different offset (i.e., audio server started at different instants), but the audio link between the two instances of the network audio software start when the first packet is sent from one of the two instances. Referring to the formalism of Sect. 3.7, let the transmitter and receiver skew be β_1 and β_2 respectively. If the two devices run at exactly the same frequency then $\beta_1 = \beta_2$, thus the offset is

$$\rho = \xi_1(t) - \xi_2(t) + T_1(0) - T_2(0). \tag{5.2}$$

As far as the random processes $\xi_1(t)$ and $\xi_2(t)$ are bounded the offset is bounded as well. As explained in Sect. 3.7, a bounded offset can be compensated for by buffering at the receiving end.

In general, unfortunately, the assumption $\beta_1 = \beta_2$ is not justified in reality. As a result the offset diverges with time. To avoid this from happening the two frequencies should be matched somehow. There is no access to the hardware clock of the two remote devices, thus the adjustment is done at the software level by means of resampling. Resampling can be introduced in a audio networking software client in two steps: a uniform resampling algorithm, allowing devices with different nominal sampling rates to exchange audio seamlessly, and adaptive resampling, allowing devices with different actual sampling rate to communicate without dropouts. By employing both, of course both advantages are achieved.

Adaptive resampling is performed by constantly adapting the resampling ratio to keep the same pace as the receiver. The data it requires to work is extracted from a software delay-locked loop (DLL), hereby introduced.

5.3.2 *Delay-Locked Loop*

A DLL is a digital implementation of a typical Phase-Locked Loop in analog electronics, i.e., a control loop employed to filter digital signals jitter and such. A reference implementation can be found in the Linux audio server JACK, later introduced in Sect. 5.4.

In analog electronics, a PLL is composed of a phase detector, a loop filter and voltage control oscillator (VCO), see Fig. 5.8.

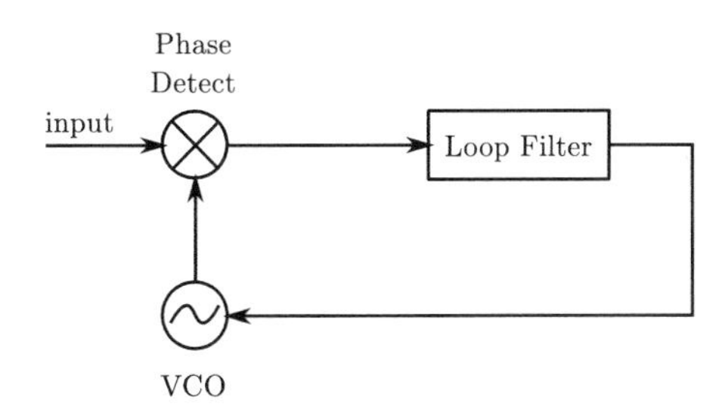

Fig. 5.8 Signal flow graph of a PLL

Let $\theta_i(t)$ be the input signal phase and $\theta_o(t)$ the VCO output signal phase. The output of the phase detector is, thus,

$$v_d = K_d \cdot (\theta_i - \theta_o) \tag{5.3}$$

where K_d is the phase detector gain. Filtering the voltage v_d by the low-pass filter $F(s)$ removes noise and high-frequency components. The VCO frequency is determined by the filter output $v_c(t)$. The relation between the VCO input and its output frequency deviation from a central value is

$$\Delta\omega = K_o \cdot v_c \tag{5.4}$$

where K_o is the VCO gain. Given, the differential relation between phase and frequency,

$$\frac{\partial\theta_o}{\partial t} = K_o \cdot v_c, \tag{5.5}$$

stands true. Its Laplace transform is

$$L\left[\frac{\partial\theta_o(t)}{\partial t}\right] = s \cdot \theta_o(s) = K_o \cdot V_c(s), \tag{5.6}$$

i.e., the phase of the VCO output is proportional to $\int v_c(t)dt$. By also transforming in the Laplace domain the following

$$V_d(s) = K_d \cdot [\theta_i(s) - \theta_o(s)] \tag{5.7}$$

$$V_c(s) = F(s) \cdot V_d(s) \tag{5.8}$$

the filter output is

$$V_c(s) = \frac{s \cdot K_d \cdot F(s) \cdot \theta_i(s)}{s + K_o \cdot K_d \cdot F(s)} = \frac{s \cdot \theta_i(s)}{K_o} \cdot H(s) \tag{5.9}$$

where $H(s)$ is the closed-loop transfer function. By substituting the low-pass filter Laplace transfer function

$$F(s) = \frac{1 + \tau_2 \cdot s}{1 + \tau_1 \cdot s} \tag{5.10}$$

in 5.9, the closed-loop transfer function can be rewritten as

$$H(s) = \frac{K_d \cdot F(s)}{\frac{s}{K_o} + K_d \cdot F(s)} = \frac{\frac{K_o \cdot K_d (\tau_2 \cdot s + 1)}{\tau_1}}{s^2 + s \cdot \left(\frac{K_o \cdot K_d \cdot \tau_2}{\tau_1}\right) + \frac{K_o \cdot K_d}{\tau_1}}. \tag{5.11}$$

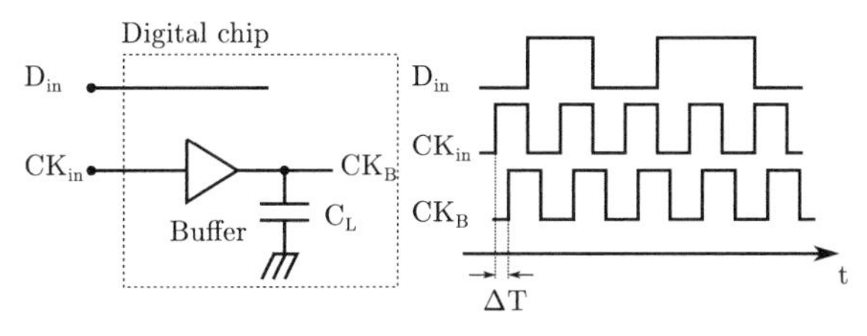

Fig. 5.9 Delay introduced by buffering and tracks parasitic components in a IC require reconstructing the clock signal

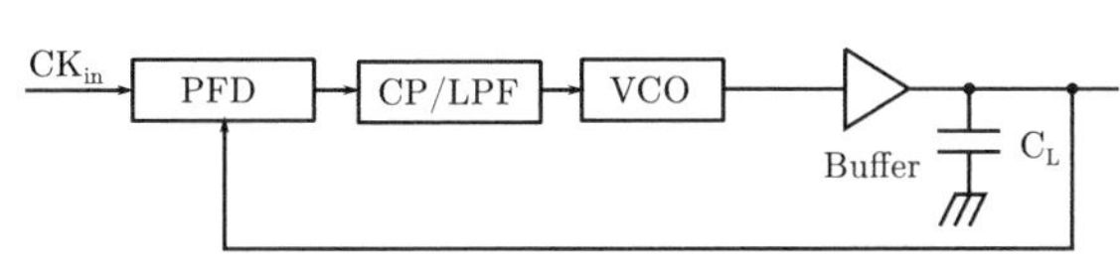

Fig. 5.10 A PLL can be introduced to remove clock signal delay or jitter

The closed-loop transfer function is thus a second-order one, and can be simply rewritten as

$$H(s) = \frac{2 \cdot \xi \cdot \omega_n \cdot s + \omega_n^2}{s^2 + 2 \cdot \xi \cdot \omega_n \cdot s + \omega_n^2}. \tag{5.12}$$

The design of PLLs for clock skew reduction is well documented [12]. Its use is necessary to remove the delay introduced by different propagation path for the clock signal through copper tracks in ICs. In Fig. 5.9 the clock signal CK_B at the input of a gate is delayed with respect from the input data D_{in} and the original clock CK_{in}. By introducing a PLL in the clock path the delay or jitter between CK_{in} and CK_b is removed by the feedback control, as shown in Fig. 5.10.

Similarly, in the digital domain, a feedback filter can be employed to filter out jitter in timestamps and estimate the periodicity between those. Since the closed-loop transfer function of a PLL is a simple second-order recursive system, it can be implemented as a IIR, shown in Fig. 5.11. Such a filter is employed in JACK [13] to estimate the current and next period timestamps as well as the period time. The filter coefficients are computed following the equations

$$\omega = \frac{2\pi B}{F_s} \tag{5.13}$$

$$a = 0 \tag{5.14}$$

$$b = \sqrt{2}\omega \tag{5.15}$$

$$c = \omega^2 \tag{5.16}$$

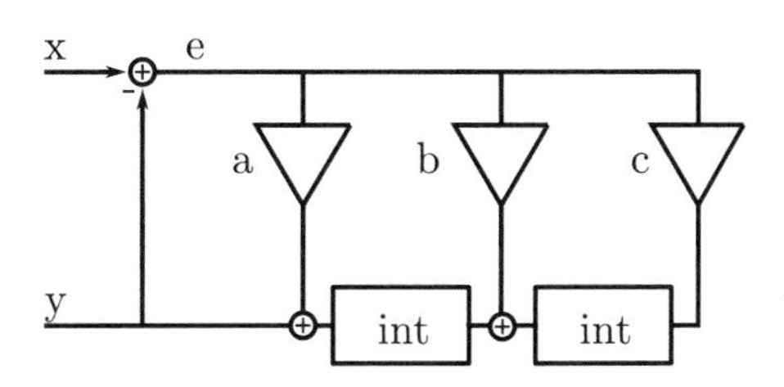

Fig. 5.11 SFG of a second-order DLL

where F_s is the sampling frequency and B the desired bandwidth.

Although the acronym is spelled as delay-locked loop [13], the literature suggests that DLL are implemented based on delay lines (see, e.g., [14]), while the DLL in JACK is more strictly related a PLL, employing a second-order filter and a means to obtain an error signal.

5.3.3 Unsupervised Resampling Strategy

A first approach to clock synchronization based on adaptive resampling is described. This approach is easily implemented and in many cases proves sufficient for the purpose. Let $\hat{F}_1(t)$ and $\hat{F}_2(t)$ be the DLL estimate of clock frequencies F_1 and F_2 at instant t. If the relation $\hat{F}_1(t) < \hat{F}_2(t)$ stands true for a sufficient amount of time, the second device goes overrun after some time. To solve the problem, the receiving end must resample according to a ratio, calculated between the two estimated clock as

$$R_r = \frac{\hat{F}_1}{\hat{F}_2} \cdot \frac{\frac{P_2}{F_02}}{\frac{P_1}{F_01}} \cdot R_u \tag{5.17}$$

where P_1, P_2 are the period size at both ends, and F_{01}, F_{02} are the nominal sampling rates at both ends, while

$$R_u = F_{02}/F_{01} \tag{5.18}$$

is the uniform resampling ratio, if the two nominal frequencies differ. Some of these parameters (P_2, F_{02}) are obtained from the transmitter when negotiating the communication, while the $\hat{F}_2$ is estimated and added by the transmitter at each period that is sent with the packet header. The packet header, thus increases from 12 bytes to 16. With typical periods of 128 frames of 16 bits (i.e., 512 bytes), the overhead increase is negligible. The $\hat{F}_1$ is estimated at the receiving end at each cycle. Both frequency estimates are calculated from JACK DLL period estimate.

This approach has been implemented in the NMP software client jacktrip. Referring to the ring buffer formalism introduced in 3.7, recovery from underruns is done in jacktrip by spinning back the read pointer of $m = 1$ slots, while with overruns the read pointer is moved forward by $m = N/2 - 1$ slots. Both solutions may not prove efficient, i.e., they depart from the ideal case of Eq. 3.7, however, they require no assumptions on the skew. In other words, by moving the read pointer backward or forward (depending on the case to recover from) by $m = N/2 - 1$ slots, an assumption is made that the over-/under-run condition occurred due to clock skew. If, on the other hand, an underrun occurred because of delayed or lost packets, but the clocks are skewed so that $F_1 > F_2$, without further occurrence of packet loss or delay, placing the read pointer close to the write pointer will reduce the time to the next overrun. The approach of the original jacktrip implementation was to make no assumptions regarding clock skew, and the behavior has been retained for the modified version as well.

The adaptive resampling mechanism requires to read an arbitrary number of samples, that are fed to the resampler and then sent to JACK for playback. For this reason the original jacktrip `RingBuffer` class has been replaced by the JACK API `RingBuffer` class. Resampling has been implemented by making use of *zita-resampler*,[9] an open-source library by Fons Adriaensen implementing polyphase filtering for audio resampling. This library provides notable features compared to other open-source libraries such as libsamplerate. Its computational cost, compared to libsamplerate is lower, even at the highest quality setting.

Of specific interest is the `VResampler` class in Zita-resampler, that allows arbitrary ratios $1/16 \leq r \leq 64$ for the resampling factor. The algorithm that is employed is a polyphase filter performing a constant bandwidth resampling in the spectral domain. Let

- F_{in}, F_{out} be the input and output sample rates,
- F_{min} the lower of the two,
- F_{lcm} their lowest common multiple,
- $b = F_{\text{lcm}}/F_{\text{in}}$,
- $a = F_{\text{lcm}}/F_{\text{out}}$,

While an ideal resampler would perform interpolation and decimation of integer factors, the resampler exploits b FIR filters in polyphase fashion. All these filters have the same frequency response, but different delays that correspond to the relative position in time of the input and output samples. The FIR filters are approximation of ideal anti-aliasing and anti-imaging filters, thus a trade-off between computational cost and aliasing is done by targeting the resampler at between the common audio sample rates (44.1, 48, 88.2, 96, 192 kHz), and considering that consequently frequency response errors and aliasing will occur only above the upper limit of the audible frequency range. Given this assumption, a trade-off is made by dimensioning the polyphase filter to reach an attenuation of 60 dB at the Nyquist frequency. As a result, aliasing is $\leq$–110 dB for all the range 0–20 kHz at sample rates of 44100 or 48000 kHz.

The approach has been implemented as part of *jacktrip*, and is currently hosted at the project repository[10] as a branch of the 1.1.0 release. The resampling algorithm has been implemented together with some heuristic rules for overrun and underrun management, and the receiver ring buffer API has been totally rewritten to allow sample by sample reading, for use with the resampler. Both uniform [15] and adaptive resampling are employed. The open-source library *zita-resampler*,[11] has been used for that. The library implements polyphase filtering for audio resampling and provides notable features compared to other open-source libraries such as *libsamplerate*. Its

[9]http://kokkinizita.linuxaudio.org/linuxaudio/zita-resampler/resampler.html.

[10]https://code.google.com/p/jacktrip/.

[11]http://kokkinizita.linuxaudio.org/linuxaudio/zita-resampler/resampler.html.

computational cost, compared to libsamplerate is lower, even at the highest quality setting.

Of specific interest is the `VResampler` class in zita-resampler, that allows arbitrary ratios $1/16 \leq r \leq 64$ for the resampling factor. The algorithm that is employed is a polyphase FIR filter performing a constant bandwidth resampling in the spectral domain. The FIR filters approximate ideal antialiasing and anti-imaging filters, and a trade-off between computational cost and aliasing is done by targeting the resampler at common audio sample rates (44.1, 48, 88.2, 96, 192 kHz), and dimensioning the polyphase filter to reach an attenuation of 60 dB at the Nyquist frequency.

5.3.4 Evaluation

The approach has been validated through Python simulations. An audio networking simulator was developed as part of WeMUST. The simulator is multithreaded, with separate threads for the transmitter and the receiver audio devices, the timebase, and the network. The network takes the data from the transmitter and delays each packet of a random value, following a uniform or a normal distribution. Multicast is allowed by the simulator, e.g., more than one receiver is allowed. The clocks of the transmitting and receiving devices can change along the simulation to emulate clock skew. In Fig. 5.12 two simulations are shown with and without resampling (unsupervised approach). In both cases the devices start at a 48 kHz clock, and the transmitter clock is later modified to obtain a positive and then negative skew. The period at both ends is 64 samples and the receiver buffer consists of eight slots. At the beginning the buffer is prefilled with zeros, in order to begin with the read and write pointer separated by half the buffer size samples. When no resampling is adopted, the pipeline (i.e., the buffer size at the receiver) is shown to progressively increase when the producer device has a higher samplerate, up to the point where buffer overrun occurs. Viceversa, when the producer device has a lower samplerate the buffer overruns. Network jitter introduces some flickering of the pipeline values. On the other hand, with adaptive resampling, the skew is corrected, thus, besides some flickering due to network jitter whenever the skew rapidly changes, the buffer size keeps constant.

5.3.5 Feedback Loop Resampling Strategy

A different approach, applying similar techniques to those presented in [16] to the networked audio case has been tested in WeMUST.

Referring to Fig. 3.6, let i be the i-th period to be transmitted, $a(i)$ the activation delay of the audio transmission client, $n(i)$ the network delay for the ith packet, $r(i)$ the time instant when the packet is received, referred to the receiver time base. The period i is copied from the hardware to the software after the transmitter's audio card

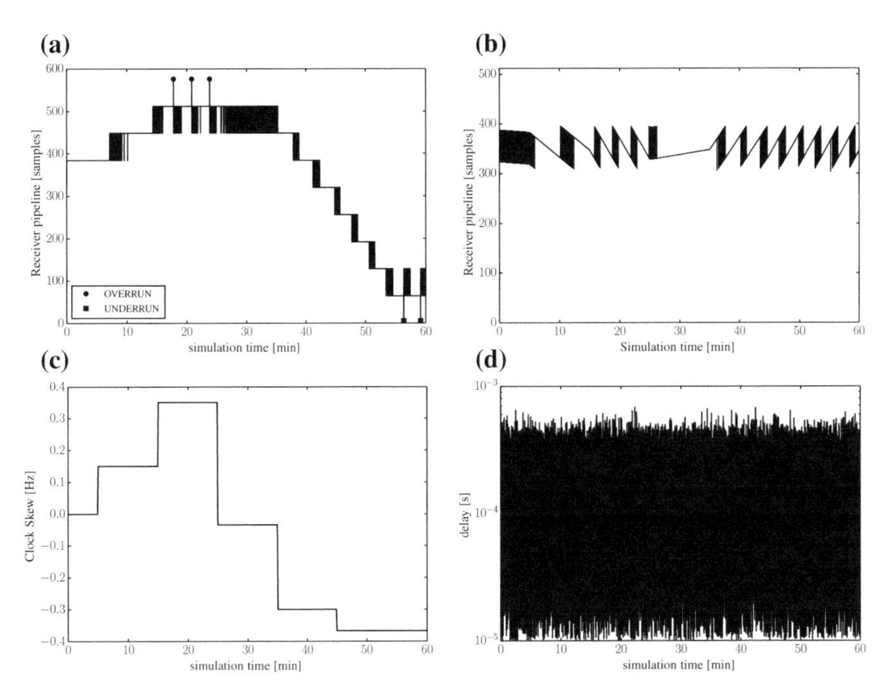

Fig. 5.12 Simulation of 1 h audio networking session, without resampling (**a**) and with adaptive resampling (unsupervised approach) (**b**). The clock skew is varied during the simulation (**c**), and the communication is subject to jitter following a normal distribution, as in (**d**)

IRQ. In the figure, the IRQ time instant is denoted as $x(i)$ and is referred to the receiver time base. Clearly, $r(i) = x(i) + a(i) + n(i)$. Unfortunately $x(i)$ is unknown and it impossible to correctly evaluate in the networked audio scenario. Furthermore, since the transmitting and receiving ends have totally unrelated time bases, it is no use to get this timestamp and transmit it with the period. Let $h(i) = r(i) - a(i) = x(i) + n(i)$ the received time minus the activation delay, and let $\mathbf{g}(\cdot)$ be the DLL filtering operator (as in [13]). If the network delay is divided into a slowly time-varying component $\bar{n}$ and a stochastic component $\nu(i)$ modeling the jitter, i.e., $n(i) = \bar{n} + \nu(i)$, a properly designed DLL is able to filter out the jitter component, i.e., the filtered signal is thus

$$\hat{h}(i)\mathbf{g}(h(i)) = x(i) + \bar{n}. \tag{5.19}$$

A continuous piecewise linear function $W(t)$ can be defined, by knowing the number of samples per period and interpolating between the $\hat{h}(i)$ period instants, that maps the number of samples sent from the transmitter at any given receiver time instant t. Similarly, the number of samples the receiver resampler consumed at a given time instant is denoted as $R(t)$.

The goal of this synchronization approach is to keep the difference $W(t) - R(t)$ constant to a predefined value, which can be seen as a *pipeline* storing transmitted samples yet to consume. In this formalism an underrun occurs when the $R(t)$ reaches the $W(t)$, i.e., no more data is available to consume, and an overrun occurs when the difference gets larger then the pipeline (which is finite). The end-to-end latency depends only on the pipeline length and the average network delay $\bar{n}$. By changing the resampling ratio according to the mechanism later described, the pipeline keeps constant and the end-to-end latency varies only with the slowly time-varying $\bar{n}$. Any wired local area network that is well below its maximum throughput should satisfy the hypothesis of a slowly time-varying $\bar{n}$. Wireless networks can satisfy the requirement if the coverage is sufficient, the number of stations to access the medium is relatively low, fading and interferences do not increase significantly the number of retransmissions per packet. A further condition to avoid dropouts is that the pipeline size must be larger than $max(n) - \bar{n}$.

In [16], adaptive resampling is detailed for two software processes on the same machine that exchange audio subject to different rates, periods, and offset. Similarly to Eq. (1) in [16], the error at any time at the receiver side can be defined as

$$e(t) = W(t) - R(t) + d_r - \Delta \tag{5.20}$$

with d_r being the resampler delay in samples and δ being the target value for latency, i.e. ,the safety buffer size. Following [16], it can be shown that the error function can be evaluated according to

$$e(t) = [W(\hat{h}(i-1)) - R(t_J)] + d_A + d_r - \Delta \tag{5.21}$$

where t_J is the last JACK activation time (i.e., audio card interrupt at the receiver side) and

$$d_A = W(t_J) = [W(\hat{h}(i)) - W(\hat{h}(i-1))] \cdot \frac{t_J - \hat{h}(i-1)}{\hat{h}(i) - \hat{h}(i-1)}. \tag{5.22}$$

Once calculated, the error is low-pass filtered, to reduce high-frequency noise content on the estimate that would otherwise manifest as a phase modulation component. Finally, it undergoes a proportional-integral (PI) control as the one used in the JACK DLL, with weights w_1 and w_2, so that the resampling correction factor has the form

$$R_c = 1 - (w_1 e(j) + w_2 \sum_{j=0}^{J} e(j)), \tag{5.23}$$

where the error function $e(\cdot)$ is evaluated at discrete time steps j up to the current JACK cycle J. The correction factor is employed to modify the resampling ratio following

$$R_r = R_c \cdot R_u \tag{5.24}$$

where R_u is the uniform resampling ratio defined in Eq. 5.18.

This approach has been implemented in *zita-njbridge*,[12] an open-source application for multichannel audio networking also implementing IPv6 addressing and allowing multicast transmission also on wireless networks. This implementation employs zita-resampler for the adaptive resampling as well.

The computational cost of this approach is higher with respect to the unsupervised one. However, the approach does not rely on estimates of clocks, and is more robustness to dropouts. The choice between the two approaches depends, thus, on the characteristics of the link (dropout probability, computational resources available on the target platform, etc.).

5.4 Real-Time Audio in WeMUST

Audio networking and networked performance software solutions have been largely discussed along the book. Most of these are based on three operating systems, GNU/Linux, Mac OS, Microsoft Windows. All three present APIs for audio capture, playback, and processing, namely Advanced Linux sound architecture (ALSA), Core Audio and Windows Audio Session API (WASAPI). Other APIs are available to provide more advanced functionalities, e.g., Audio Stream Input/Output (ASIO) under Windows or Jack Connection Kit (JACK)[13] under Linux. Additionally cross-platform APIs, such as PortAudio, may provide compatibility between different operating systems. The choice of the operating system and the audio APIs is not trivial, as each have different characteristics and advantages. Windows is employed in the LOLA project to take advantage of the efficient proprietary drivers available for several video cameras, in order to have minimum latency. Mac OS X is employed by a large number of experimental composers and musicians for it is considered a very stable platform for music computing. For this reason, all the music computing platforms (Pure Data, Max/MSP, Supercollider) are well supported and NMP projects, such as Soundwire, ported their software to this platform as well. Finally, GNU/Linux is the only open-sourced operating system among the three, and it is the only one to be portable on embedded devices. Furthermore, it is highly configurable, well suited to be modified for experimental purposes and the functioning of its audio drivers and audio servers are well known and documented. Many NMP softwares are available for GNU/Linux, including Ultra Video, Ultragrid, jacktrip, and more. It has also been chosen as the reference platform in WeMUST, making it possible to build NMP applications with embedded portable devices. In this context, the choice of the audio API is rather straightforward: JACK is quite established as the reference framework and API for audio and signal processing applications development, providing the concept of deadline, daisy chaining of audio processes, and timing information as those required for clock synchronization (see Sect. 5.3.2). Alternatives discarded are the Open Sound System (OSS) and PortAudio. OSS has been deprecated by Linux

---

[12]http://kokkinizita.linuxaudio.org/linuxaudio/.

[13]www.jackaudio.org.

kernel developers and its APIs are now only provided inside ALSA as an emulation layer for legacy software compatibility. PortAudio provides cross-platform portability providing an intermediate layer between audio applications and several APIs. However, JACK is available too on all the three major operating systems.

An overview on digital audio capture and playback has been provided in Sect. 3.3. The sound card A/D gathers samples at a certain rate imposed by the clock and generates an interrupt every time a *period* (power of 2-sized chunk of the audio buffer) is ready to be passed to the software. The interrupt is serviced by the operating system, but actual audio processing and buffer handling may be deferred depending on thread priority. The interrupt servicing routine triggers a cascade of function calls, proceeding through a number of software layers.

JACK does not totally replace ALSA, instead it relies on its kernel drivers to take control of the audio hardware. ALSA is the currently maintained Linux kernel audio driver and API implementation for low-level audio on Linux, replacing the legacy driver OSS. ALSA manages the hardware and provides the buffering mechanism for audio input and output. It only allows one client to take exclusive control of the hardware.

JACK is a set of APIs and an audio server. It enables real-time processing within its active clients, i.e., audio applications that make use of JACK APIs. JACK clients consist of real-time threads for audio processing and other nonreal-time threads for control, user input, graphic user interface, etc. JACK can request to ALSA exclusive access to the hardware but allows any number of audio clients to exchange audio to and from the hardware and among themselves. All the clients are synchronized to the audio card interrupts. When JACK is started all the audio parameters (bit-depth, sample rate, period size, etc.) must be specified. These parameters cannot change dynamically, and all the audio clients must work at the specified samplerate and period size (although internally they may resample).

To initialize audio processing, a user application can register one its functions to be executed at each JACK cycle (i.e., every period time) via the following callback.

```
int jack_set_process_callback (jack_client_t *client,
                                JackProcessCallback
                                process_callback,
                                void *arg)
```

Listing 5.2 JACK process callback method

After the audio engine starts, the `process_callback` function is executed at each period time and the internal time function $T(t)$ is started. On single-core computing architectures all the audio processing threads of the JACK registered clients are executed in a serial fashion and their input and output ports can be daisy-chained. On multicore processors, the audio thread can execute in parallel when no dependencies are present (i.e., a thread input does not depend on the output of another thread). All the audio threads must complete processing in the current time-frame in order avoid audio dropouts. This is called synchronous mode in Jack. Clients can be internal (those strictly related to JACK inherent functions, e.g., audio capture and playback) or external, i.e., user applications, as seen in Fig. 5.13.

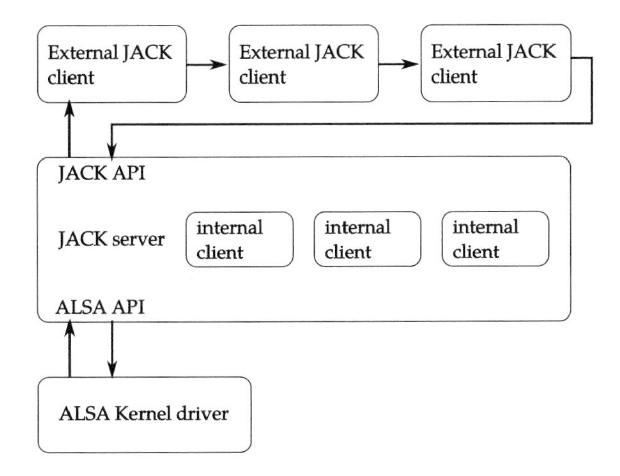

Fig. 5.13 Internal and external clients in JACK

In ALSA and JACK audio samples are handled in arrays, called *periods*. Each period consists of an interleaved sequence of left/right samples (in the case of stereo audio), called *frames*. Period sizes can range from 32 to 1024 and above frames. Shorter periods allow for reduced latency, but they increase the computational overhead as the number of interrupts increase. The latency is also related to the buffer size, i.e., the number of periods that are allocated for exchange of audio data to and from the audio card. A minimum of two periods must be allocated in the buffer, in order to perform a *ping-pong* strategy: while the hardware writes a period, the other one is passed to the software for reading, and viceversa. Figure 5.14 clarifies the concept related to buffering. Note that two periods are always in use by the software or the hardware for passing audio data. By allocating more periods in the buffer the process of exchanging audio to and from the hardware is safer, however extra latency is added. The input–output latency is

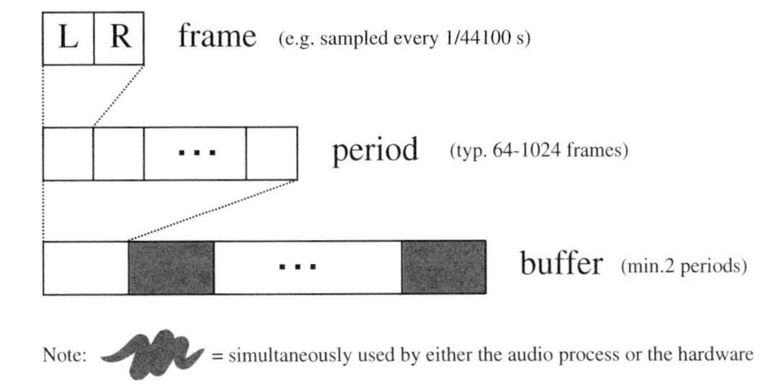

Fig. 5.14 Naming convention used in the Linux audio community

$$D_{i/o} = 2 \cdot N \times \frac{P}{F_s}, \tag{5.25}$$

where N is the number of buffers (with a minimum of 2, i.e., the ping-pong case), P the period size—normally of 2^i samples. To this, the delay introduced by any algorithms (zero only if audio is sent back to the output or no components with memory are employed in the processing) must be added. Interrupts from the audio card signal the presence of a new period to read and require a new period to play. For each period, JACK reads the input period and writes back an output period with audio to play back. The reading, processing and writing back must be done in the time between two audio card interrupts, i.e., a *period time*. In order to make audio work properly, JACK is called whenever an interrupt from the audio card is generated and allows its clients to be scheduled by the operating system. In Linux, several scheduling strategies are implemented. The completely fair scheduling (CFS) algorithm aims at maximizing CPU utilization and process reactivity [17, 18]. For this reason it is the default scheduler for user processes. However, for real-time processes a set of fixed priority queues (`SCHED_FIFO`) exists that can have higher priority than regular CFS-scheduled threads. JACK real-time threads can be scheduled in a `SCHED_FIFO` queue, to preempt other user processes that do not have strict timing requirements. Jack clients run a real-time DSP thread under `SCHED_FIFO` and other nonreal-time threads under the CFS (`SCHED_OTHER`) scheduler. Every Jack client RT thread gets a priority lower than that of Jack but higher than most userspace processes, thus, Jack can preempt clients that do not release the CPU when the deadline is reached (new interrupt). Every time this happens a dropout can be noticed, since the period returned to the hardware does not contain proper data. The scheduling mechanism is reported in Fig. 5.15.

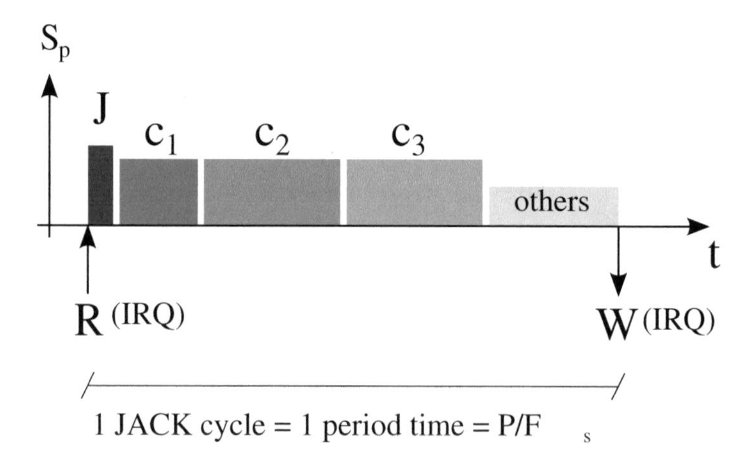

Fig. 5.15 Audio processes with scheduling priority S_p scheduled on a single-thread processor. The JACK audio routine (J) is executed first, due to its higher priority, at the interrupt of the audio card. Three JACK clients (c_1, c_2, c_3) follow. Their scheduling order is imposed by JACK according the dependencies imposed by their routing. For the sake of simplicity only the read from the audio card (R) and the write (W) are shown

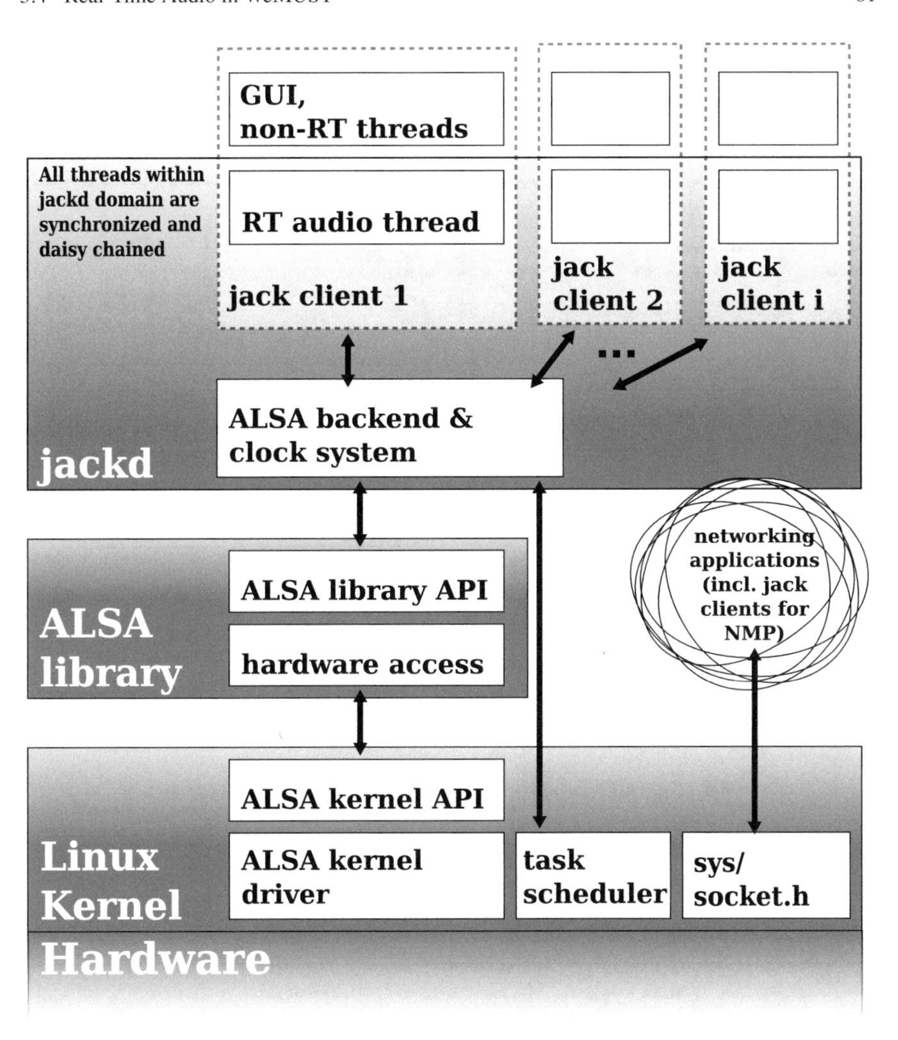

Fig. 5.16 Overview of the audio software stack with a generic Linux use case. JACK is employed to run several audio threads

The audio software stack is reported in Fig. 5.16. Figure 5.17 reports all the components of ALSA and OSS (Open Sound Server), the legacy sound server on Linux systems.

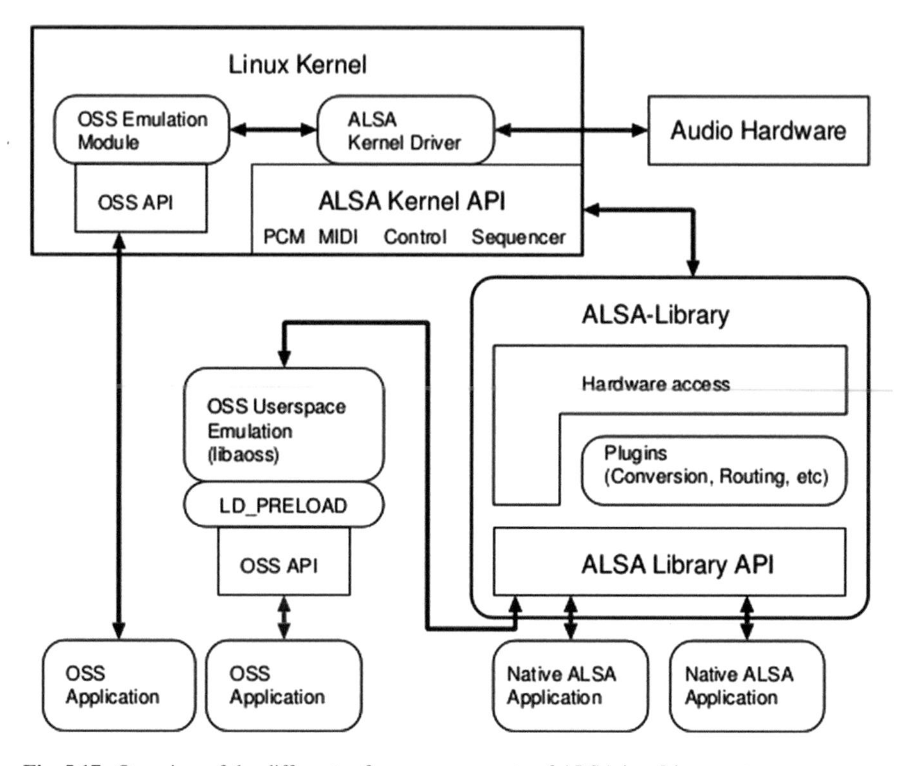

Fig. 5.17 Overview of the different software components of ALSA in a Linux system

5.5 Audio Networking Software in WeMUST

The JACK audio server natively supports audio networking through the following modules: *netjack1* and *netjack2*. In alternative two external JACK clients, jacktrip [4] and zita-njbridge are available. Each one of these has its own advantages and disadvantages, making it desirable for certain features but inadequate for some uses. For instance, netjack1 supports uncompressed audio streams as well as CELT compression. Typical CELT coding delay is very low, but stands between 5 to 22.5 ms. An interesting feature in netjack2 is the availability of device discovery, which however does not follow any specific protocol, and is implemented using multicast for announcement as done in many other protocols (see Sect. 5.2). All this software is bidirectional with the only exception of zita-njbridge, which requires instantiating both a transmitting and a receiving process for bidirectional audio networking. Synchrony is handled in a master-slave approach for netjack1 and netjack2, i.e., the slave does not synchronize to the local audio card, but to the master period. Natively, jacktrip does not support synchronization, but the adaptive resampling approach described in Sect. 5.3.3 has been implemented by the authors in a branch of the stable version of the software. Furthermore, zita-njbridge implements the adaptive

Table 5.1 Comparison of different audio networking applications in JACK

	Compression	Link	Auto discovery	Synchrony
netjack1	Uncompressed or CELT	Master-slave	None	Master
netjack2	Uncompressed	Master-slave	Yes	Master
Jacktrip	Uncompressed	Peer-2-peer	None	None
Zita-njbridge	Uncompressed	Monodirectional	None	Adaptive resampling

resampling approach described in Sect. 5.3.5. Table 5.1 highlights the main features for each one.

Clearly, each application have its advantages and disadvantages. One valuable feature of jacktrip is its cross-platform portability. Being based on Qt libraries[14] it is supported by the three major operating systems. Other desired features in the context of NMP is the availability of a device discovery mechanism, and bidirectional links. Multicast transmission, supported only by zita-njbridge is equally useful in some context.

Jacktrip has been considered in WeMUST as the target software, although the others have been tested and experimented as well. Jacktrip is released as an open-source software[15] and has been improved during the development of WeMUST [15], branching the original source code of the v1.1.0 release. Either the Debian stable repository version (Jacktrip v1.0.5) and the modified one have been tested and employed in the WeMUST project. The v1.1.0 is more CPU-intensive than v1.0.5, mainly because of the v1.1.0 architecture reworking. When all the devices in the network share the same audio parameters, v1.0.5 is viable, while the modified code is needed for heterogeneous settings due to the addition of resampling. Jacktrip is employed to acquire audio from the hardware and send it to a remote client through the network. It also allows to receive audio from the same client. This could be a processed version of the original signal, or a monitor mix of several sources.

5.6 ARM Platforms for Wireless Audio Networking

As outlined in Chap. 2 most if not all NMP works have been performed based on general purpose computing platforms. Desktop and laptop computers have been employed to transmit audio and video signals acquired from a stage to a remote end. In other cases, performers share data from their laptop on a stage. As far as personal computing can go with miniaturization and improvements can be done in HCI, the inherent format of the laptop or the personal computer is bound to the size

[14] http://www.qt.io/.

[15] https://github.com/jcacerec/jacktrip.

of a screen and cannot shrink below a certain size or accomodate certain usages.[16] For this reason, the musical performance is limited by the portability of the object and the physical interaction with the instrument is limited by the input interface. The NIME academic community is often concerned with conceiving and prototyping physical interfaces to improve human–machine interaction in the performative context. Most of the efforts go towards Personal Area Networks, with , e.g., wearable sensors connected to a central device through short-range communication technologies. However, miniaturization enables the whole system to be within human range and acquire not only sensory data but also acoustic signals. Following, the advent of mobile computing, fast prototyping and development boards based on embedded processors such as the ARM family, were adopted and are inexpensive. For this reason to push portability further, wireless connectivity can be coupled with embedded physical computing. Several projects make use of two successful embedded ARM boards, designed by the Raspberry Pi foundation[17] and the BeagleBoard.org foundation.[18] Several platforms are now available, each with its own advantages and disadvantages.

The BeagleBoard xM (BBxM) is taken as a reference, featuring rather high performances compared to the others and having integrated audio input and output. The BBxM is an open hardware platform, featuring a 1 GHz ARM Cortex-A8 core (Texas Instruments DM 3730), 512MB RAM and many peripherals generally found on personal computers or mini-PCs. One important aspect is the availability of input and output line sockets, and a capable audio codec[19] IC (Texas Instruments TPS 95650). The input analog stage is lacking a preamplifier to adapt to microphone signal levels. More recent and widespread platforms, such as the Beagleboard Black by the same foundation, or the Raspberry Pi, are similar in the core architecture, but are targeted to different application, and either lack the onboard audio codec or lack the audio input. For this reason, some works addressed this need by adding an external audio codec or USB sound card [19, 20]. In WeMUST, for instance, portability and ease of connection is a key goal, hence the BBxM, with its on-board audio connections and codec has been preferred over other platforms.

Key components available on the platform are the Ethernet and USB controllers, the audio codec, the RS232 serial interface for debugging and the microSD slot which stores the bootloader and the operating system. The availability of the Ethernet controller is fundamental in prototyping stages, to access the system reliably via a remote command console. The USB connection is necessary to add WiFi capabilities or to attach commodity devices.

The BBxM are driven by a full operating system stack. In WeMUST a customized Debian Linux system has been proposed, named WeMUST-OS, including all the tools

---

[16] Unless some concepts related to screen morphology and rigidity are surpassed.

[17] https://www.raspberrypi.org/.

[18] http://beagleboard.org/.

[19] Any integrated circuit including audio ADC and DAC is often called a *codec* in the engineering jargon.

needed for employing it for wireless NMP. The image for BBxM and all the software tools are available at the A3LAB research group web page.[20]

5.6.1 *Power Management*

Multimedia applications can often be power constrained. Power management is required for battery-powered applications, and there are use cases, such as that described in Sect. 6.1 which require to run on battery. To avoid audio glitches and achieve maximum performance the DM3730 core must not resort to dynamic frequency scaling and for best performance it should be clocked at the highest frequency, i.e., 1 GHz during audio packets exchange or processing. However, while waiting for a connection, the core can be clocked at a lower frequency. This functionality, called frequency scaling, is implemented in the `cpufreq_governor` Linux kernel module, generally referred to as *scaling governor*, and is easily accessible from the `sysfs` virtual file system interface. The scaling governor enables different operating modes, such as *ondemand*, *performance*, *powersave*. While the former dynamically scales frequency according to a load-balancing algorithm, the latter two enable maximum and minimum core frequency. By taking advantage of the `sysfs` interface WeMUST-tools scripts enable to save power while not transmitting audio. As musical performances need to be set up some time prior to the due time (even hours before), saving power while waiting for the performance to start enables to increase battery life.

5.6.2 *On-Board Audio*

Embedded hardware may or may not feature an audio codec on board. In very general terms the audio codec is connected to the main processors by a dedicated serial bus. An on board codec generally connects to the application processor through a serial connection such as the Inter-IC Sound bus (I^2S), which is a three wire serial bus with master and slave devices. With external codecs USB is a very common solution as it is able to transport the large amount of the data (Firewire, a well-established serial bus for audio cards, is seldom featured in embedded hardware). The application processor has a buffer related to the serial line connected to the codec which is necessary to temporarily store data to and from it.

While other embedded audio projects employ external USB audio interfaces [21] or custom tailored I2S (Inter-IC Sound) audio codecs [20], the BBxM has the advantage of hosting an on-board audio codec, the TPS95650. This is a companion chip supporting operation of OMAP3 and other OMAP3-compatible ICs (such as the DM3730 application processor). It provides a codec for high-quality multimedia

[20]http://a3lab.dii.univpm.it/research/wemust.

streams and a voice codec, for lower quality, low latency voice communication. The audio codec is connected to one of the DM3730 serial lines, McBSP2 (Multichannel Buffered Serial Port). The McBSP2 has a large buffer, to avoid glitches in multimedia applications. In [15] the viable input-output latency was shown to be 9.3 ms. This practical limitation is given by the processor and could be reduced with other platforms, with a benefit for NMP applications.

The choice of the buffer size determines the computational resources devoted to buffer exchange, interrupts servicing, etc. The lower the period size, the higher the overhead. For instance, with the BBxM ARM core running at maximum speed and JACK, the minimum buffer size that do not incur into glitches is 64 frames. Lowering the period size to 32 incur into audible glitches.

One important lack of the TPS95650 is the absence of a microphone preamp, forcing the user to use an external preamplifier. 9 V battery-powered ones are a available for smartphones and similar devices which provide +48 V *phantom* power, accept XLR balanced inputs and adapt it to the 3.5 mm jack input of the BBxM, however the availability of a microphone preamp and phantom power is required with most microphone capsules necessary for high-quality audio.

5.6.3 A/D and D/A Latency and CPU Overhead

As mentioned previously, the specific hardware and software configuration reported above introduces a input-output latency of 9.3 ms without affecting the audio quality with audible glitches. This is given by the need for 2 64-frames audio input buffers and 5 64-frames audio output buffers at 48 kHz. However, when transmitting audio to other peers, some ring buffering is implemented inside Jacktrip and zita-njbridge, to avoid network jitter. This buffering is necessary to avoid glitches due to network issues, but adds some latency. With good networking condition, four buffers can be allocated, with an average delay of two buffers (the buffer pointer tries to keep in the middle of the ring buffer to avoid both overruns and underruns). Tests with two BBxM acting as peers, with the signal following a round trip from one to the other and back via Jacktrip, the average input-output latency is approximately 15.7 ms, with 6.4 ms given by the ring buffer (two receive ring buffer, one for each trip) and the remainder 9.3 ms by the BBxM hardware buffering. No resampling is performed and both BBxM are set to work at 48 kHz.

Another important aspect is the CPU load of such a configuration. The shorter the period size, the shorter the time to perform all operations (interrupts servicing, buffer handling, data encapsulation in 802.11 or 802.3 frames, just to name a few). This results in a higher overhead and a higher average CPU load. Table 5.2 reports some average CPU load data depending on the audio parameters for a system running JACK, sending and receiving audio packets through Jacktrip. The table is reported to highlight some nonlinearities of such systems. It may be noted, e.g., at a constant frame size, an increase of the sample rates should yield a proportional increase in the overall CPU load. This is verified with a frame size of 128, where the CPU load

Table 5.2 Average CPU load evaluated for different sample rates and period sizes

		32 kHz (%)	44.1 kHz (%)	48 kHz (%)
128	Jack	13	19	21
	jacktrip	16	21	23
64	Jack	21	30	46
	jacktrip	22	33	46
32	Jack	89	78	94
	jacktrip	–	–	–

Please note that with a period size of 32 frames, Jacktrip has not been run since JACK already required most of the CPU for synchronizing to the audio card interrupts

increases from 32 to 44.1 kHz of 1.37 as expected, and similarly for the 44.1 and the 48 kHz rates, the increase is 1.1 (1.08 expected). However, with frame size 64, the measured CPU load increases of 1.53 from 32 to 44.1 kHz and of 1.39 from 44.1 to 48 kHz. In other words, by halving the frame size, the overall CPU load increase factor is 1.4, 1.5, and 2.1 for the three different sample rates. With frame size 32 JACK even obtains a lower CPU load at 44.1 kHz, possibly because of an improved hardware or kernel driver optimization for that sample rate. It is important to note that the upper bound for glitch-less audio operation is below the 100 % CPU load threshold, due to the nonoptimal nature of the Linux scheduling algorithms (in terms of deadline constraining) and the presence of kernel preemption and hardware interrupts. Kernel network layer critical sections and network card interrupts servicing, that are important for audio packet transmission can affect performance. In the current case, for instance, the interrupt routine for the wireless or wired LAN controller, may steal time to the audio process interrupt routine managing the TPS95650.

5.6.4 Wireless Transceivers

Embedded hardware may or may not include an on board wireless transceiver. This, generally interfaces with the application processor by means of a bus. often communicate through a Serial Peripheral Interface (SPI), a master-slave full duplex 4 wires protocol, able to reach large bandwidths for inter-IC communication in a serial fashion. External transceivers require to be connected through a bus such as USB. Publicly available transceiver for external use through USB are those for WiFi IEEE 802.11 communication or Bluetooth. Exceptionally, some sub-1 GHz or ZigBee transceivers are available for debug and development purposes as USB devices for personal computers. A the time of writing, reliable 802.11ac Linux kernel device drivers able to obtain high data rates are not available. However 802.11a/b/g and 802.11n transceivers are well supported and provide large data rates. In the development of WeMUST-OS, e.g., a recent kernel has been employed that includes several wireless chipsets drivers.[21] Among these, the RT5370 (driver: rt2800usb) and RTL8191SU (driver: rtl8712u) chipsets have been employed extensively.

While driving a on-board or USB-connected wireless chipset is the most straightforward solution to connectivity within WeMUST, wired Ethernet connection facilitates the development and may be considered as a fallback solution.

5.7 Enabling Wireless Transmission in Critical Contexts

In previous works by the authors [3, 15, 22], short-range indoor transmission is taken in consideration as the target scenario. As a result, commercial USB WiFi chipsets have been employed. These generally feature omnidirectional 2.4 GHz/5 GHz patch antennas or larger dipole antennas and only rarely there is the option of mounting a different antenna.

In Sect. 6.1 an outdoor live performance is reported which took place in a critical environment: the sea. The conditions imposed by this and other peculiar NMP cases make a carefully designed wireless transmission critical for the outcome of the performance. Specifically, critical factors such as long distances, the presence of water, hot sand, trees or other sources of diffraction, moving obstacles and metal objects required the RF link to be conceived differently. The problem can be posed in terms of link budget. Assuming that the choice of the transceivers (and their parameters such as sensitivity and transmission power) is done, the first option to consider is adopting carefully crafted antennas, to obtain higher antenna gains. If possible, directional antennas can be employed to increase the range and reduce scattering and reflections. The obvious shortcoming is to keep the antennas in line-of-sight as much as possible. Additionally, to increase robustness, redundancy can be exploited, by creating a different RF link for each audio networking device. Redundancy can be done, e.g., by employing separate frequency bands for each link. This, although increases the BOM (Bill Of Materials) for a live performance (each band requires a receiver tuned to that band), guarantees greater reliability. Finally, it is generally suggested to use the 5 GHz ISM band instead of the 2.4 GHz one, as it is less crowded and antenna patterns are narrower compared to a 2.4 GHz antenna of the same size. Similarly, by employing Sub-1 GHz radio technologies, most potential interferers are dodged. Obviously a field test to look for potential interferers is mandatory, in order to avoid channels occupied by competing networks with strong SNRs (Signal to Noise Ratio).

Given the experimental nature of NMP, the operation of tuning the links, measuring bandwidths, scanning the frequency band of interest for interferers can be complex, and tools to ease the work and improve performance reliability are helpful. To facilitate the deployment and design of the performance, Ethernet to wireless bridges, such as those from the SXT family from Mikrotik,[22] can be employed. These devices

[21] http://wireless.kernel.org/.

[22] http://www.mikrotik.com/.

are configurable wireless stations/access points, comprising a 10/100 Ethernet port, a directional antenna, driven by a high-power RF amplifier, and a microprocessor. Their size is slightly bigger than that of a typical embedded ARM board, with the antenna surface having a diameter of 140 mm. Configuration is done by software from a personal computer through a browser and a large set of features are available, including routing, bridging, and acting as an access point. Wired and wireless interfaces are totally configurable and the RF link can be set up using standard 802.11 modes or even proprietary high-throughput protocols. The SXT can also scan an area for networks and monitor SNR and link quality during transmission. A robust NMP can be deployed by connecting each audio node with an SXT by employing the Ethernet port, instead of using a local bus as it would be done with USB adapters. This also reduces scheduling issues due to USB driver implementations and competing devices accessing the USB bus.

The solution has been assessed during the development of WeMUST, to judge on the viability of the approach and measure feasible SNR margins for live usage. Preliminary experiments have been conducted in an empty indoor space of 19×9 m. This space, although not ideal, was free from neighboring wireless networks, obstacles or reflective surfaces and stands as a good reference for best-case transmission, with SNR values very close to the maximum ones achievable. Tests were performed setting up three separate 802.11a links, on three adjacent frequency bands, of 20 MHz width each inside the 5 GHz ISM band. The antennas and BBxM were put on the side opposite to the mixing PC, i.e., at 19 m. The SXT can automatically set for the best data rate, however in these favorable conditions, the data rate always stays at the maximum 54 Mbps and the SNR is quite stable at an average 110 ± 2 dB with maximum Tx power. For what concerns the used bandwidth, in these conditions, when sending stereo signals in both directions with 32-bit samples at 48 kHz, the SXT monitoring utility measured Tx/Rx rates of 1.7 Mbps/1.7 Mbps including the overhead added by Jacktrip and 802.11.

In the target environment the SNR can drop dramatically from the 110 dB figure reported above, thus to properly design the RF link parameters it is necessary to collect statistics on link quality and find SNR margins that guarantee a very low rate of audio glitches occurrence. In order to gain insight on this, a set of tests has been performed reducing the SNR by decreasing the transmission power. The interest in this context is on the rate of audio periods lost, i.e., those that did not arrive in time or were not received at all. The period loss rate (PLR) index is, thus, introduced, which measures the rate of lost periods over the total periods in the time unit. To evaluate the PLR sine tones with offset were generated from the BBxM, sent to the mixing PC and recorded there for later analysis by software. When the receiver is missing a period it writes zeros to the output file, for easy detection of lost periods. The test results are reported in Fig. 5.18. As the Figure shows the salient aspect is the sudden increase of lost packets when the link SNR falls below 38 dB. Unexpectedly, there is also a slight increase in the packet loss at very high SNR values due to nonlinear effects at the RF frontend of the SXT (most prominently intermodulation and input signal saturation). It is suggested, thus, that a best-case scenario for evaluating transmission

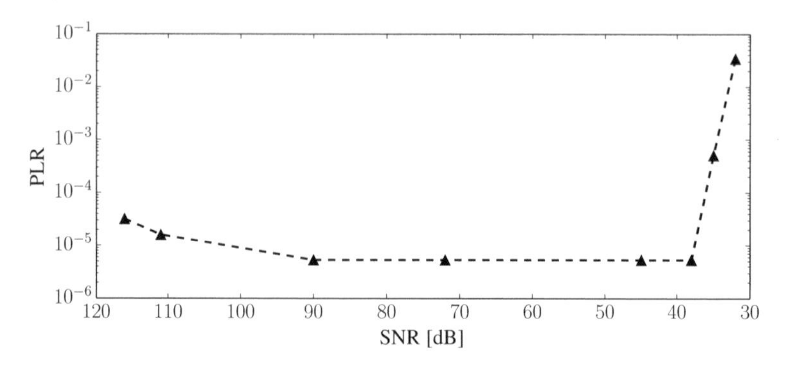

Fig. 5.18 Period Loss Rate tests at different SNR. All tests have been conducted at the same Data Rate (36 Mbps) but decreasing Tx power, thus decreasing the SNR

performance is not the first one reported above, with a very high 110 ± 2 dB SNR, but a lower Tx power configuration, which achieves an SNR between 90 and 40 dB.

In the attempt to reduce the SNR required to decrease the packet loss, several strategies have been implemented, including:

- varying the number of transmission attempts on a missing ACK,
- introduce redundancy in Jacktrip, i.e., sending systematically a packet more than once.

Unfortunately neither solution proves useful. A redundancy in Jacktrip even increases the PLR, due to the doubled bandwidth. Sending data over UDP devoids from the chance to ask for retransmission, but TCP is not a solution due to the large increase in latency and its unpredictability.

When collecting signals from N remote ends into one mixing PC, for the sake of reliability and robustness, it is suggested to create a link for each end. However, this N-by-N solution requires $2N$ SXT, plus one Ethernet network switch. Another feasible setup needs $N + 1$ SXT only. In this case one of the SXT acts as Access Point (AP) and the other as Stations, however all communication is performed on one channel only, requiring enough bandwidth for transmission of all the signals. Furthermore, there is risk of collisions and increase of transmission delays due to errors at one end. In other words, the isolation provided by the more robust solution allows the performance not to be affected by a broken link. PLR tests have been performed for a 3-to-1 case, analog to the Waterfront case study described in Sect. 6.1, together with a multichannel 1-to-1 case involving the same amount of traffic. Figure 5.19 provides PLR results for two data rates and three different configurations. The baseline is the first case where two bidirectional stereo channels are exchanged between two nodes. The other two cases share the same overall bandwidth handled by the node at the mixing PC end, i.e., 4.6 and 9.2 Mbps (excluding overhead). In the 3-by-1 scenario, these cases are implemented:

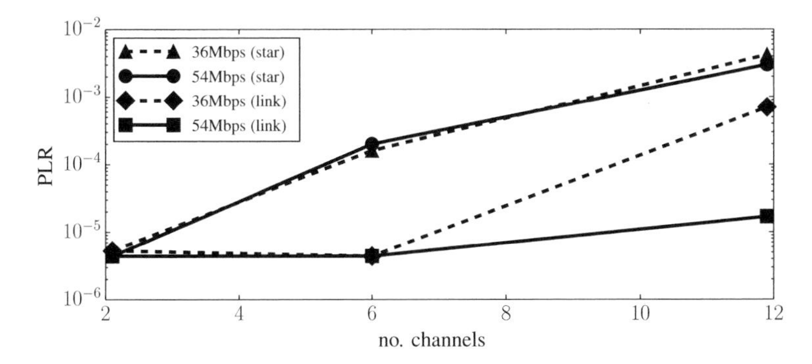

Fig. 5.19 Period Loss Rate tests with different multichannel configurations. Please note, values of 0.74×10^{-6} (*bottom values*) mean that no period loss was observed during an interval of 60 min

- three nodes sending stereo signals to a mixing PC and the latter sending a stereo mix back to them, for a total of six bidirectional streams (average 5.2 Mbps uplink, 5.2 Mbps downlink)
- three nodes sending 4-channels signals to a mixing PC and the latter sending a 4-channel mix back to them, for a total of 12 bidirectional streams (average 9.9 Mbps uplink, 9.9 Mbps downlink).

In the 1-to-1 scenario 6 or 12 audio channels are sent in both directions (with total 5 and 9.7 Mbps uplink and downlink).

As the Figure shows the PLR increases with the number of channels for both tests, however, as expected, the 1-to-1 case has a lower PLR due to the decreased chance of collisions.

For the sake of comparison, it must be considered that with the N-by-N solution the PLR stays equal to the baseline for all the connections. Reducing the PLR for multichannel or N-by-1 requires further effort which is left to future works, however viable alternatives may exploit the proprietary TDMA protocols available with the SXT.

All the results presented in this Section are for single-hop networks. Should a longer signal chain be devised requiring multiple hops, further evaluation should be done following the procedures in [23].

References

1. Reuter J (2014) Case study: building an out of the box Raspberry Pi modular synthesizer. In: Linux Audio Conference (LAC2014). Karlsruhe, Germany
2. Letz S, Denoux S, Orlarey Y (2014) Audio rendering/processing and control ubiquity? a solution built using faust dynamic compiler and JACK/NetJack. In: Joint Internation Computer Music Conference and Sound and Music Computing (ICMC+SMC14). Greece, September, Athens, p 1518

3. Gabrielli L, Squartini S, Piazza F (2013) Advancements and performance analysis on the wireless music studio (WeMUST) framework. In: AES 134th convention, May 2013
4. Cáceres J-P, Chafe C (2010) Jacktrip: under the hood of an engine for network audio. J New music Res 39(3):183–187
5. Cheshire S, Krochmal M (2013) RFC 6762: Multicast DNS. In: Internet Engineering Task Force (IETF) standard
6. CheshireS, Krochmal M (2013) RFC 6763: DNS-based service discovery. In: Internet Engineering Task Force (IETF) standard
7. Freed A, Schmeder A (2009) Features and future of Open Sound Control version 1.1 for NIME. In: Proceedings of the conference on New Interfaces for Musical Expression (NIME), Pittsburgh, PA, USA
8. Eales A, Foss R (2012) Service discovery using open sound control. In: Audio engineering society convention, vol 133. Audio Engineering Society
9. Bowen N, Reeder D (2014) Mobile phones as ubiquitous instruments: towards standardizing performance data on the network. In: Joint international computer music conference and sound and music computing (ICMC+SMC2014). Greece, Athens
10. Ogborn D (2012) Espgrid: a protocol for participatory electronic ensemble performance. In: Audio engineering society convention, vol 133. Audio Engineering Society
11. Weibel H, Heinzmann S (2011) Media clock synchronization based on PTP. In: Audio Engineering society conference: 44th international conference: audio networking. Audio Engineering Society
12. Razavi B (2002) Design of analog CMOS integrated circuits. Tata McGraw-Hill Education
13. Adriaensen F (2005) Using a DLL to filter time. In: Linux audio conference
14. Yang C-KK (2003) Delay-locked loops-an overview. Phase-Locking in High-Peformance Systems. Wiley-IEEE Press, New York, pp 13–22
15. Gabrielli L, Bussolotto M, Squartini S (2014) Reducing the latency in live music transmission with the BeagleBoard xM through resampling. In: Proceedings of the European embedded design in education and research conference, Milan, Italy
16. Adriaensen F (2012) Controlling adaptive resampling. In: Linux audio conference, Stanford, USA
17. Molnar I Modular scheduler core and completely fair scheduler CFS. Linux-Kernel mailing list
18. Wong C, Tan I, Kumari R, Lam J, Fun W (2008) Fairness and interactive performance of o(1) and CFS linux kernel schedulers. In: International symposium on information technology, 2008. ITSim 2008, Aug 2008, vol 4, pp 1–8
19. Topliss J, Zappi V, McPherson A (2014) Latency performance for real-time audio on Beaglebone Black. In: Linux audio conference (LAC2014). Karlsruhe, Germany
20. Meier F, Fink M, Zölzer U (2014) The Jamberry-a stand-alone device for networked music performance based on the Raspberry Pi. In: Linux audio conference, Karlsruhe
21. Berdahl E, Ju W (2011) Satellite CCRMA: a musical interaction and sound synthesis platform. In: International conference on New Interfaces for Musical Expression (NIME), Oslo, Norway, 30 May–1 June 2011
22. Gabrielli L, Squartini S, Principi E, Piazza F (2012) Networked beagleboards for wireless music applications. In: EDERC 2012, September 2012
23. Rahman Siddique M, Kamruzzaman J, Hossain M. An analytical approach for voice capacity estimation over WiFi network using ITU-T E-model. IEEE Trans Multimedia 16(2):360–372

Chapter 6
Applications

Abstract Based on the developments described in the previous chapter, use cases are reported hereby for wireless audio networking in networked music performance. All the technical and logistical details are reported and stand as a reference for future improvement, development of similar systems, and for composers to plan on similar performances and artists to conceive multimedia art installations that employ wireless audio networking. Suggestions to exploit further features given by wireless audio networking are given.

Keywords Live performance · Energy supply · Outdoor wireless audio networking · Art installation

6.1 Waterfront: A Networked Music Performance Experience

One of the outcomes of wireless audio networking, is the ability to bring audio networking and NMP in outdoor and challenging environments. As a demonstration, during the development of the WeMUST project, conducted by the authors, a live outdoor performance has been envisioned to showcase the framework and receive a feedback from trained musicians. The performance was conceived for *Acusmatiq*, a live electronic music festival, taking place in Ancona, Italy, in the magnificent frontage of the *Mole Vanvitelliana*, an eighteenth-century pentagonal building constructed on an artificial island close to the port of Ancona and other historical buildings. Nowadays, the Mole has decks for motor boats and sailboats, metal obstacles (such as metal lift docks) and wireless networks for maritime purposes, which can endanger the wireless audio transmissions. The goal was to have acoustic musicians play on separate boats at the frontage of the Mole, each with its own equipment to stream their signal to the land, and receive back a mix of the ensemble playing. The audience would stand on the land and be free to roam, exploring the acoustic space and observe the musicians from different vantage points. The performance has been called *Waterfront*.

L. Gabrielli and S. Squartini, *Wireless Networked Music Performance*,
SpringerBriefs in Electrical and Computer Engineering,
DOI 10.1007/978-981-10-0335-6_6

Fig. 6.1 Satellite view of the Mole Vanviteliana and the sorrounding area. *1* Direction console with master PC, *2* land antennas standing on the Mole fortified wall, *3* double bass, *4* saxophone, *5* vocalist, *6* audience

Figure 6.1 reports the satellite view of the area. The maximum transmission distance is that between the land antennas and the double-bass player (2 and 3 in Fig. 6.1), i.e., 90 m. The shortest is that between the land antennas and the vocalist, (2 and 5), i.e., 30 m. The pier (6), is 54 by 14 m wide. The antennas are placed on top of the fortified wall, at 3 m height, to leave space for the audience to roam along the pier.

After a careful selection of possible artistic proposals and performers, the choice converged on a trio performance of *Solo* by Karlheinz Stockhausen. The piece, dated 1966, explores on a solo performance (hence the name), augmented by feedback tape delays which selectively loop some of the acoustic material to enrich the scene and provide a seemingly ensemble experience. As opposed to this, with Waterfront, the three musicians are granted a moment of solo performance each, but join finally into an impromptu ensemble where each instrument gains its space competitively, reinforced by the distorted recurrence of the feedback delays. The whole performance lasts approximately 50 min.

The original piece required four technical assistants to manually drive the delay taps, gains, and filters according to six different variations suggested by Stockhausen himself in the form of notated scores. In this piece the signal processing was conducted digitally with *SOLO nr.19*, an iPad application. Each of the musicians had its own iOS device running the aforementioned application, which captured the instrument signal and supplied the BBxM with the signal. This also compensated for the lack of a microphone preamp on the BBxM, since the iPad outputs a line signal, compatible with the BBxM line input. The iOS devices add latency to the signal chain. This was discussed with the musicians, who could test with different setups and latency settings and they finally decided to have a slightly larger latency but perform with a less cumbersome equipment. The processed version of the signal is insensitive to the low processing delay added by the device, since the delays imposed by the Solo score can be up to several seconds long. As a side note, consider the architecture of the WeMUST framework: in order to reduce the use of hardware devices a countdown and a *click* to synchronize musicians can be supplied from the MPC, together with all the signal processing required by Stockhausen's score (porting of the signal processing in Solo exist for CSound and other languages). This would also reduce the latency added by the iOS device. The musicians however felt more comfortable carrying their own iOS device, which also provides visual feedback.

A video of the performance is provided at the A3LAB research group WeMUST page.[1]

6.1.1 Energy Supply

To evaluate the energy requirements of the system current consumptions have been measured for both the BBxM and the SXT. Current probes have been used to measure current consumption, while the voltage supplied by the regulated power supply is constant at 5 V for the BBxM and 12 V for the SXT, independent of the output current. Power requirements have been measured during operation, and are reported in Table 6.1 as a function of the Data Rate and the Tx Power.

The musicians have been supplied with a BBxM and an SXT each. To power this equipment commercial-grade 12 V 4.5Ah rechargeable lead-acid batteries have been used, enabling sustained operation. The voltage must be regulated to fit the requirements of the equipment. A switching voltage regulator, the LM2596 (adjustable output version), has been employed. This allows up to 2 A output current (without heat sink) and input voltage in the range 4.5–40 V. The efficiency depends on the output voltage. For the case at hand ($V_{IN} = 12$ V, $V_{OUT} = 5$ V) the efficiency is

[1] http://a3lab.dii.univpm.it/research/wemust.

Table 6.1 Current consumption for the SXT and the BBxM under different operating modes

Status	TX power (dBm)	SXT (mA) at 24 V	BBxM (mA) at 5 V
Searching	17	115	465
Waiting	17	125	
Transmitting 24 Mbps	0	143	600
	12	145	
	17	148	
Transmitting 54 Mbps	0	134	
	12	136	
	17	138	

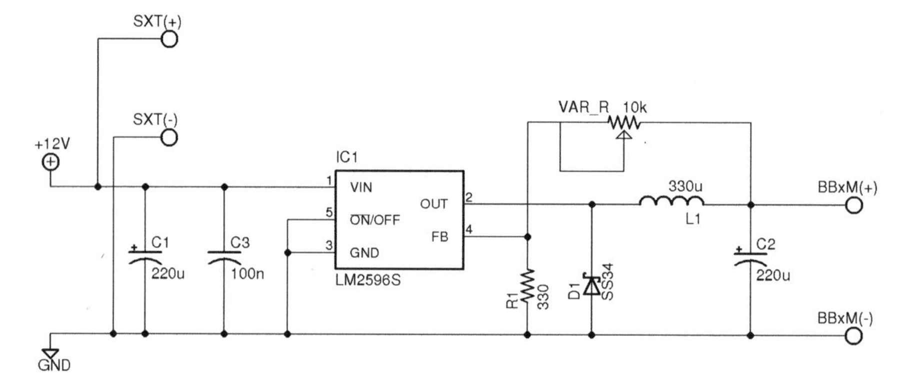

Fig. 6.2 Schematic diagram of the battery supply circuit

80 %. The typical quiescent current is 5 mA at 25 C. To meet more stringent power constraints, the quiescent current can be reduced to 80 μA, by switching the component off (toggling pin 5, shown in Fig. 6.2). The SXT have an internal regulator and can be directly supplied with 12 V.

During indoor transmission tests the battery life was evaluated. As expected the batteries performed slightly different from each other, however, the minimum battery life documented in the tests was of 3 h of continuous transmission plus 1 h of waiting. Given the figures shown in Table 6.1, the total power required by the system while transmitting is 6.3 W.

6.1.2 Audio Signal Routing and Monitoring

The signal routing is depicted in Fig. 6.3. As mentioned beforehand, the BBxM acquires the line signal from the iOS device, which is used to capture the sound from the acoustic instrument, and output the original and processed version. The BBxM acquires the signal and transmits it over to the SXT, which acts as a transparent bridge

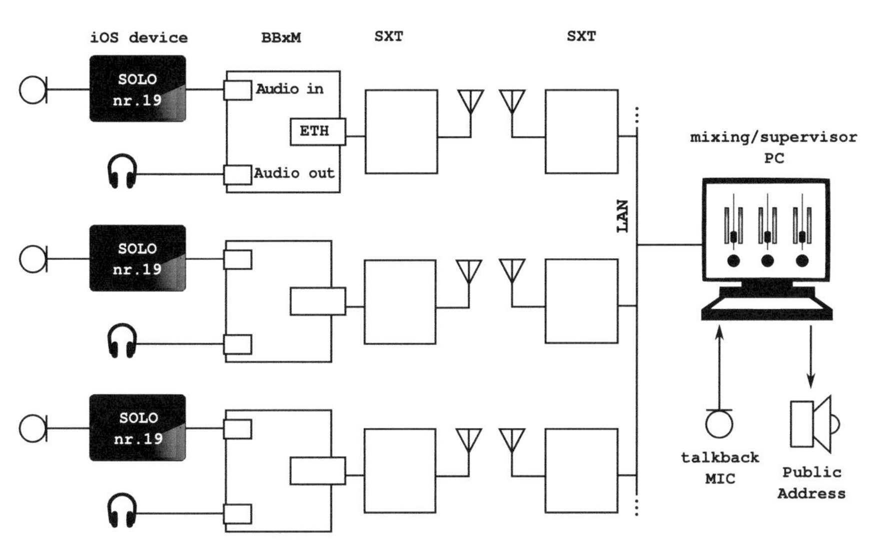

Fig. 6.3 Overview of the hardware and connections employed in Waterfront

to the mixing PC. As all the signals are acquired at the master PC, they are mixed exploiting a software mixing console. The direction could speak to the musicians with an open microphone connected to the MPC sound card. By proper routing (which can be usually automated by means of scripts and presets) the signals are mixed for the public address system and are selectively sent to the musicians, to allow each one to have its own earphone monitor mix, to ensure the best interaction experience. All the musicians experience, thus, the same round-trip time, which, with the final settings agreed for the premiere performance, was 27.4 ms, i.e., 15.7 ms (as documented in Sect. 5.6.3) plus an iOS latency of 11.7 ms (128 sample buffers, includes some proprietary microphone signal preprocessing). As stated above, the latency added by the iOS devices can be reduced by moving all the signal processing inside a JACK client in the MPC.

During the performance audio glitches were partly masked by employing a rudimentary error concealment mechanism available in Jacktrip, i.e., buffer repeating. By offline analysis on the performance recording, three glitches were spotted in approximately 50 min, up to the expectations from previous transmission tests.

During the Waterfront premiere, a MacBook Pro was used as the MPC, running *JackOSX*, Jacktrip, *qjackctl*, and *AULab*. Equivalently, on a Linux machine, JACK, Jacktrip, *qjackctl*, and *jack-mixer* can be employed to obtain similar features. For the sake of a stable performance, a Linux machine was employed, running *wemust-netdbg* to monitor the state of the connections. This configuration requires one technician managing audio, levels and musicians' requests and another to control the state of the connections. With the system getting more mature both audio and monitoring tools could run on the same machine.

6.2 Other Live Performance Use Cases

In Waterfront a star topology was employed, with the mixing PC collecting all the signals. However, distributed topologies are allowed by WeMUST and wireless NMP in general. Let us consider a network of audio networking nodes, with omnidirectional antennas and an open space where the musicians can roam. The audio nodes can collect audio and send it in a multicast fashion by employing zita-njbridge. Each node can, algorithmically or after human action, select whether to receive and process or playback the audio data sent by each one of the other musicians. Signal processing can be performed at each node. The processing involved with Solo, e.g., could be moved to the audio nodes in the form of a music computing software (e.g., Puredata, Supercollider, etc.), without largely affecting the CPU load (the processing is based on delay lines) but reducing the latency. Similarly, as mentioned beforehand, the mixing PC could process each one of the incoming audio signals. This solution would free the musicians from the use of a tablet device, and could sustain computation of algorithms of higher complexity that require increased computational power.

One downside of employing a headless platform is the lack of a visual cue or screen to provide musicians with information, a visual metronome, etc. On the other hand, modern tablets have good computational resources and large displays but lack openness of development platforms, the availability of general purpose Input/Output lines for sensor data acquisition, and are weighted down with general purpose software of no use in this context.

In Waterfront no control data was exchanged. However, MIDI or OSC data can be exchanged along the network, e.g., to remotely trigger events or control one of the endpoints running some processing or synthesis algorithm. Another use for control data typically found in networked laptop orchestras [1, 2], is a broadcast of status messages, control parameters, or data exchange for multi-agent or distributed architectures. All these options are allowed by WeMUST, as any device connected to the network can transparently send or receive data to and from any other device. Connections and devices on the network can dynamically change. Any device can be addressed by its IP addresses or hostname, and these can be discovered by use of device discovery protocols, such as SABy.

One last noteworthy feature enabled by WeMUST is to ensure the musicians the ability to synchronously play a score, by sending a metronomic *click* from the MPC or a conductor device to the endpoints.

6.3 Installations

Media art installations often make use of real-time audio and video processing, or playback. There are several use cases that can benefit from low latency wireless audio networking. Very often art installations are based on audio/video material played back repeatedly for the whole installation duration. Often, art pieces are associated with a soundwork that is played by a loudspeaker hidden in the proximity of the artwork. When two or more devices need synchronous playback wireless networking can provide a feasible solution, either by multicast transmission of signals from a central device and employing the clock synchronization mechanisms described in Sect. 5.3 to maintain synchronization, or by synchronizing audio playback with regular triggers from a master (and provide a means to ensure jitter removal from these triggers). Figure 6.4 reports a schematic image of the art installation scenario, with a number of nodes and a master providing synchronization or multimedia content by multicast transmission. The master node may also be equipped with sensors to track the position of one or more users, and convey this information to the WeMUST nodes that may process the audio content based on the user's position. Similarly, the nodes may acquire sensory data to send to the master for processing and fusion. On the basis of this, the master may send processed audio to the nodes. Proximity sensors close to art pieces can be also useful to gather statistics regarding user interest and report them to the master. Some interactive art installations also capture and process or display audio and video content gathered from the installation room. This can be done with wireless nodes and the information can be delivered to other nodes for display or processing in other areas of the venue. These scenarios may provide additional creative inputs to artists and foster new interactive art installations. It must also be

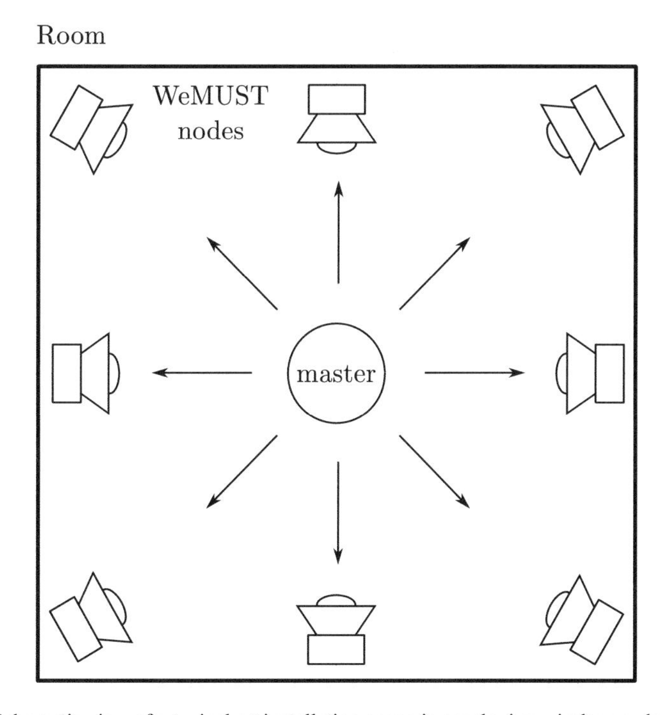

Fig. 6.4 Schematic view of a typical art installation scenario employing wireless nodes for control and audio exchange

noted that avoiding running lengthy audio cables can decrease costs and facilitate temporary installations.

Distributed audio and data acquisition strategies can be drawn from the emerging field of Wireless Acoustic Sensor Networks (WASN); thus wireless nodes can be seen as wireless audio sensors in a sensory network. A steadily increasing corpus of literature is emerging in the field, reported with more detail in Chap. 7.

References

1. Blain M (2013) Issues in instrumental design: the ontological problem (opportunity?) of liveness for a laptop ensemble. J Music Technol Educ 6(2):191–206
2. Dahl L (2012) Wicked problems and design considerations in composing for laptop orchestra. In: International conference on new interfaces for musical expression (NIME)

Chapter 7
Conclusions and Future Directions

Abstract As reported in Chap. 2, at the moment of writing wireless networking is still to be fully exploited for networked music performance. Some attempts to use IEEE 802.11 networking for laptop orchestras are reported, but no technical achievements are described and performances have not been evaluated. On the contrary, it seems that there is very little confidence with wireless networking for NMP and most authors relied on commodity hardware and had limited usage of wireless networking. Wireless audio networking and wireless NMP are, thus, open fields for exploration. The works covered in this book, mainly contributed by the authors, show the feasibility of the concept with a relatively low engineering effort for demonstration and live usage, but many issues are open for thorough exploitation of the wireless medium and improved integration with existing software and composition frameworks. The possibilities opened by the availability of networking technologies able to sustain wireless audio transmission are vast and are worth investigating. Indeed, wireless NMP deserves attention as it provides an exceptional means to value indoor and outdoor spaces, monuments or parks, and opens up new possibilities of expression for composers and performers. The development of wireless technologies may improve on everyday usability of music technologies such as controllers, effect processors, or studio recording and mixing devices. More details on current issues and possible avenues for research in wireless NMP are provided in this chapter.

Keywords Future developments · Psychoacoustic · Software interoperability · Wireless acoustic sensor networks

7.1 Software and Interoperability in NMP

There are attempts in the academic music community to provide protocols for music networking, session handling, and discovery. Open Sound Control has been proposed for service discovery [1], networking and sharing of musical objects [2], and queries.[1] Others proposed connection and payload formats for audio networking (e.g., JACK,

[1] Schmeder-Wright, MrRay, OSNIP, Minuit, QLAB, libmapper, oscit, to name a few.

L. Gabrielli and S. Squartini, *Wireless Networked Music Performance*,
SpringerBriefs in Electrical and Computer Engineering,
DOI 10.1007/978-981-10-0335-6_7

jacktrip, and so on). As long as these efforts are sparse and no clear perspective on the future of OSC is available it is not suggested to spend more effort on a device discovery and negotiation protocol. Finally, industrial groups propose several complete protocol stacks for audio networking that generally hardly fit for NMP applications and often do not guarantee interoperability with other protocols. The publishing of the AES47 standard for interoperability and the adoption of AVB for non-audio purposes may solve the latter issue.

For what concerns the NMP software, current features are very limited. Most open software include command line applications that need some glue logic or graphical user interface to improve usability, automation, and adoption. Audio resampling is not often available and cross-platform availability is rarely provided. Video transmission is provided only in a few software applications. Among the NMP software only Linux-based applications can also run on embedded hardware, meaning that most solutions do not match requirements of portability. WeMUST-OS only supports the BeagleBoard xM as embedded platform. As time passes by, new more performing development boards will come out, which will draw the attention of users. Some of these platforms should be supported by WeMUST-OS, in order for more users to experiment with WeMUST.

A small engineering effort could make these open software applications interoperable, since they are mostly based on similar concepts, payload formats, etc. Interoperability would be achieved by simply agreeing on:

- a metadata structure that includes all the relevant audio parameters, to assess feasibility of the audio link, instantiate a resampler if needed, allocate the proper audio buffer size, etc.;
- a set of header and payload structures for audio exchange: most audio software (netsend, jacktrip, AUnetsend, netjack, etc.) exchange audio in similar ways, but the packet header, numeric data representation, and interleaving are slightly different;
- a negotiation protocol to exchange the metadata, negotiate audio link features, and establish a connection;
- a session control protocol; it can be drawn from existing protocols but it needs adoption by all NMP audio software.

Another option would be the adoption of the AES67 interoperability standard [3]. Its implementation in existing software could solve the aforementioned issues. The AES67 standard also provides for a minimum set of requirements in terms of latency and audio packet size. These requirements are 48 kHz, 24-bit sampling, 48 samples per period, and 3 periods per buffer. This means 3 ms buffer time. With a maximum expected network delay within the range of 3 ms an AES67-compatible transmission could be done.

7.2 Wireless Technologies

Some protocols in the IEEE 802.11 family have been extensively tested during the development of the WeMUST project. At the physical level, some problematic aspects that can be outlined are:

- ISM 2.4 and 5 GHz bands have rather high attenuation with distance. The sub-1 GHz range would be better suited to longer transmission range. The IEEE 802.11af standard can be regarded as a good candidate for future outdoor tests, given its extended range and sufficiently large bandwidth.
- ISM 2.4 and 5 GHz bands are largely employed and the number of potential interferers is rather high in urban areas.
- Directional antennas with adaptive beamforming could be of high interest, increasing transmission range while allowing movement of the signal source.

The distributed medium access strategy of IEEE 802.11 proved adequate during tests but its limitations can be foreseen in NMP: if the number of nodes grows, redundancy is exploited or retransmissions are often requested due to large link attenuation. Advanced medium access mechanisms provided in IEEE 802.11e do not prove useful for audio networking as all audio packets have equal importance and, thus, priority. For concurrent transmission of multiple audio packets in a short time frame a time-slotting mechanism could be employed, similarly to what is done in the Linux scheduler with audio processes, which are queued and executed in a round-robin fashion. This would increase throughput by removing overhead and backoff wait time. Several TDMA (Time Domain Multiplexing) algorithms are implemented in the Mikrotik devices described in Sect. 5.7 as an alternative to standard 802.11 medium access policies that could be tested.

7.3 Joint Development with Wireless Acoustic Sensor Networks

A novel topic that is attracting the interest of the academic community is that related to distributed acoustic sensing and processing with wireless nodes, i.e., Wireless Acoustic Sensor Networks (WASN). This rapidly emerging topic is of high interest for its applicability to a large number of scenarios and application fields. Some of the technical outcomes in wireless audio networking can also be exploited in WASNs. On the other hand, the expanding number of researchers in the WASN field brings forward new concepts and developments to the field of wireless audio networking that cannot be obtained by the small wireless audio networking and wireless NMP community.

WASN developments started a few years ago [4, 5] and proved of interdisciplinary nature from the beginning, as they involve aspects of networking, wireless transmission [6], security and coding [7], digital signal processing, sampling, sparse representation, acoustic beamforming [8], data fusion, and distributed parallel processing [9]. Wireless acoustic sensors can be seen as irregular microphone arrays and thus signal processing strategies to render an even distribution of the microphone is necessary. Clock synchronization, audio fluxes coordination, and strategies to reduce unnecessary data throughput are required. Latency is often bounded by the application at hand and the available processing power. Energy efficiency is an issue, since acquiring and processing acoustic signals requires more power than conventional wireless sensor networks.

With the increasing interest in wireless acoustic sensor networks, some benefit can be foreseen to the field of wireless NMP, and wireless audio networking research may cross-fertilize with WASN research.

7.4 Psychoacoustic Research Issues in NMP

Psychoacoustic research can still offer insights into several questions, especially those related to the role of video in NMP. The importance of video signals in NMP is yet to be assessed. There is discordance on whether low latency video signals are necessary for synchronization, and whether they can improve reaction and interaction during a performance. As aligning audio and video requires audio to undergo the high latency imposed by video, it should be verified whether audio and video need to be perfectly aligned or some delay between the two can be tolerated.

Another interesting psychoacoustic topic is that related to sound source localization and spatialization in NMP. Live acoustic music offers a very rich listening experience in terms of spatialization, depending on the settings, which expose the listener to the direct sound radiated by the instruments and its blending with the room acoustic. Depending on this the listener obtains a very clear acoustic image of the space sorrounding him. On the contrary, electroacoustic live sound deployment generally devoids the experience from its real spatialization, with the sound reinforcement system providing a bland or distorted spatialization idea based on panning of the electric signal of the sound sources. Current NMP experiences provide the same experience. In the future, however, with the development of proper sensing and analysis techniques, NMP can be envisioned that transmit relevant data to feed a spatialization resynthesis technique to provide an acoustic image of the remote ensemble employing precision sound reinforcement systems.

To conclude, contemporary music composers and artists alike should provide input and test cases to conduct research. Their acceptance of wireless networking in music performance, however, is hard to foresee in the near future given its novelty. Furthermore, besides technical and usability issues there are also public health concerns related to wide usage of radiated electromagnetic fields.

References

1. Eales A, Foss R (2012) Service discovery using open sound control. In: Audio engineering society convention 133. Audio Engineering Society
2. Bowen N, Reeder D (2014) Mobile phones as ubiquitous instruments: towards standardizing performance data on the network. In: Joint international computer music conference and sound and music computing (ICMC+SMC2014), Greece, Athens
3. AES standard for audio applications of networks: high-performance streaming audio-over-IP interoperability. Audio Engineering Society, 60 East 42nd Street, New York, NY, USA, 2013
4. Akyildiz IF, Melodia T, Chowdhury KR (2006) A survey on wireless multimedia sensor networks. Comput Netw 51(4):921–960
5. Bertrand A (2011) Applications and trends in wireless acoustic sensor networks: a signal processing perspective. In: 2011 18th IEEE symposium on communications and vehicular technology in the Benelux (SCVT), pp 1–6
6. Rüngeler M, Vary P (2015) Hybrid digital-analog transmission for wireless acoustic sensor networks. Signal Process 107:164–170
7. Zahedi A, Østergaard J, Jensen SH, Bech S, Naylor P (2015) Audio coding in wireless acoustic sensor networks. Signal Process 107:141–152
8. Zeng Y, Hendriks RC (2012) Distributed delay and sum beamformer for speech enhancement in wireless sensor networks via randomized gossip. In: 2012 IEEE international conference on acoustics, speech and signal processing (ICASSP). IEEE, pp 4037–4040
9. Malhotra B, Nikolaidis I, Nascimento M et al (2008) Distributed and efficient classifiers for wireless audio-sensor networks. In: 5th international conference on networked sensing systems, INSS 2008. IEEE, pp 203–206

Appendix
Audio Networking Standards

Since NMP represents a niche field in the academic literature, it often takes advantage of existing technologies for signal transmission. Currently digital audio transmission is covered by a large number of protocols. Of particular interest for NMP applications are those standards that fall under the umbrella terms of audio over IP (AoIP) and audio over Ethernet (AoE). A number of very mature and developed protocols belong to this family, which are used in broadcasting studios, recording studios and such, for the delivery of musical content at very low latency and high reliability.

The efforts in the direction of digital audio delivery started at least 30 years ago with the ratification of the AES3 standard, also known as AES/EBU as it was developed by the Audio Engineering Society with the contribution of the European Broadcasting Union [1]. The AES3 protocol defines the synchronous transmission of a stereo PCM signal over several transmission media, and is incorporated into IEC 60958 standard. In principle, it defines a time-slotted technique for sending stereo audio data in 24-bit or 20-bit sample words. The data is uncompressed and the time slotting depends on the audio sampling rate. Allowed sampling rates are 32, 44.1, and 48 kHz. A simple representation of the time slotting employed by AES3 is provided in Fig. A.1. Commercial variants are S/PDIF (which only transmits stereo data) and ADAT optical interface. ADAT optical interface or ADAT Lightpipe introduces a larger bandwidth, allowing up to 8-channel 48 kHz 24-bit audio. Similarly, MADI introduces multichannel transmission following the AES10 standard [2], and employs coaxial or fiber optic cables. It employs asynchronous time slotting as in the previous standards, so that time slots are not synchronized to the audio sampling rate. The latest amendment to the standard allows up to 64 channels at 48 kHz. Synchronization of the frames is not done by the frame overhead, but by symbols external to the frames. All the aforementioned protocols are widely used in practice for digital transport of audio between specialized hardware and computer audio peripherals; however, they are not suitable to switched packet networks. AES47 and AES51 [3, 4], for instance, respectively, define how to pack AES3 audio data on generic asynchronous networks and specifically on Ethernet hardware. Such communication technologies are leveraged daily in the music broadcasting and distribution fields [5]:

L. Gabrielli and S. Squartini, *Wireless Networked Music Performance*,
SpringerBriefs in Electrical and Computer Engineering,
DOI 10.1007/978-981-10-0335-6

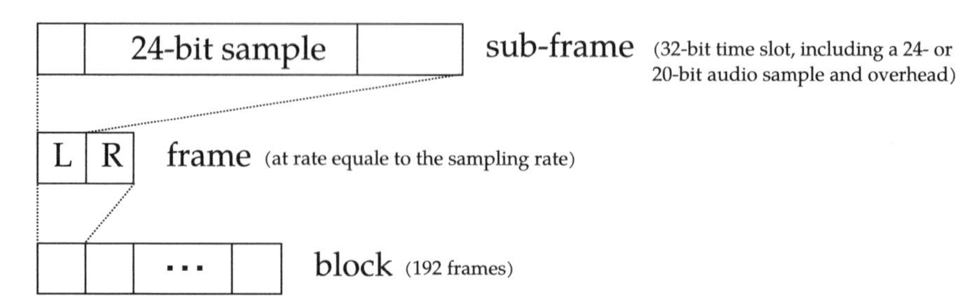

Fig. A.1 A simplified representation of the AES3 or AES/EBU data format and time multiplexing

audio-over-Ethernet technologies are widely used for the local area [6] (broadcasting studios, conservatories, universities, large rehearsal studios, arenas and sport venues, train and metropolitan area alert and voice diffusion, etc.). Commercial products from several manufacturers employ open or proprietary protocols that encapsulate audio over standard Ethernet packets, such as CobraNet and EtherSound. Other protocols encapsulate data in IP packets and are independent of the underlying communication layers. Among these IEEE 802.1BA (also known as AVB), RAVENNA, Q-LAN, LiveWire, and Dante can be counted. The former is taken as an example to enumerate several of the typical features of an audio-over-IP technology. The Dante protocol is based on UDP-encapsulated audio transmission over existing networking infrastructures such as Fast Ethernet, Gigabit Ethernet, or fiber optic IEEE 802.3 links. It also shares a master clock, synchronized between all the Dante-enabled machines using precision time protocol (PTP, standard IEEE 1588), enabling misalignment between the networked audio devices of typical 1 μs. Devices are automatically discovered and can be routed or controlled through the network, by employing a software tool provided by the manufacturer. Audio, timing, and connection traffic can coexist with an existing networking infrastructure employed for other non-audio related traffic (as it would be in a school or conservatory local area network), provided that the infrastructure allows for traffic prioritization through DiffServ code points (DSCP), reserving, thus the maximum priority to timing messages (for clock synchronization through PTP), and a medium priority to audio traffic. One advantage of employing general-purpose networking technologies is the ability of employing a personal computer as a Dante device. The protocol does not support IP networking over the Internet or over wireless networks for reasons of reliability and latency; however, it enables fiber optic links even at very distant geographical locations.

IEEE 802.1BA or AVB was originally meant for audio and video, and developed by the Audio Video Bridging Task Group of the IEEE. Later in 2012, it was renamed to Time-Sensitive Networking task group, and the scope of the standard was broadened, thus, potentially increasing the number of adopters of the protocol. AVB provides a synchronization mechanism coherent to the stringent specifications of low latency streams, a bandwidth reservation protocol, delay reduction mechanisms on IEEE 802 networks, and interoperability profiles. Although targeting IEEE 802 networks, the protocol only provides synchronization support for wireless IEEE 802.11 networks, based on OSI layer 2 primitives providing time stamping.

Given the relatively large number of proposed standards and commercial technologies, an interoperability standard for audio networking was produced recently known as AES67 [7]. This standard proposes a set of minimum features all the producer must adhere to provide interoperability between the aforementioned technologies. At the moment of writing many suppliers of audio-over-IP systems are providing an AES67 compatibility layer for interoperability with other suppliers' equipment. AVB also provides interoperability features.

To the best of our knowledge, there is no report of networked music performance employing the aforementioned audio-over-IP protocols. One reason is their restriction to local area networking. The issues introduced by the Internet or a wireless medium require specific solutions that need to be addressed properly and all the current solutions described in literature were implemented by researchers by writing software from scratch.

References

1. Specification of the Digital Audio Interface (The AES/EBU interface) Tech 3250-E, 3rd edn. European Broadcasting Union, 2004
2. AES10-2008: Recommended Practice for Digital Audio Engineering-Serial Multichannel Audio Digital Interface (MADI). Audio Engineering Society, 2008
3. IEC62365: Digital audio—Digital input-output interfacing—Transmission of digital audio over asynchronous transfer mode (ATM) networks, 2009
4. AES51: AES standard for digital audio—Digital input-output interfacing—Transmission of ATM cells over Ethernet physical layer. Audio Engineering Society, 2006
5. Rumsey F (2011) Audio in the age of digital networks. J Audio Eng Soc 59(4):244–253
6. Shuttleworth T (ed) (2011) AES Technical Committee on Network Audio Systems. Emerging Technology Trends Report
7. AES standard for audio applications of networks: high-performance streaming audio-over-IP interoperability. Audio Engineering Society, 60 East 42nd Street, New York, NY, USA, 2013

Printed in the United States
By Bookmasters